Hodder Gibson Model Practice Papers WITH ANSWERS

PLUS: Official SQA Specimen Paper & 2014 Past Paper With Answers

National 5

Mathematics

2013 Specimen Question Paper, Model Papers & 2014 Exam

This book contains the official 2013 SQA Specimen Question Paper and 2014 Exam for National 5 Mathematics, with associated SQA approved answers modified from the official marking instructions that accompany the paper.

In addition the book contains model practice papers, together with answers, plus study skills advice. These papers, some of which may include a limited number of previously published SQA questions, have been specially commissioned by Hodder Gibson, and have been written by experienced senior teachers and examiners in line with the new National 5 syllabus and assessment outlines, Spring 2013. This is not SQA material but has been devised to provide further practice for National 5 examinations in 2014 and beyond.

Hodder Gibson is grateful to the copyright holders, as credited on the final page of the Answer Section, for permission to use their material. Every effort has been made to trace the copyright holders and to obtain their permission for the use of copyright material. Hodder Gibson will be happy to receive information allowing us to rectify any error or omission in future editions.

Hachette UK's policy is to use papers that are natural, renewable and recyclable products and made from wood grown in sustainable forests. The logging and manufacturing processes are expected to conform to the environmental regulations of the country of origin.

Orders: please contact Bookpoint Ltd, 130 Park Drive, Abingdon, Oxon OX14 4SE. Telephone: (44) 01235 827720. Fax: (44) 01235 400454. Lines are open 9.00–5.00, Monday to Saturday, with a 24-hour message answering service. Visit our website at www.hoddereducation.co.uk. Hodder Gibson can be contacted direct on: Tel: 0141 848 1609; Fax: 0141 889 6315; email: hoddergibson@hodder.co.uk

This collection first published in 2014 by
Hodder Gibson, an imprint of Hodder Education,
An Hachette UK Company
2a Christie Street
Paisley PA1 1NB

BrightRED PUBLISHING Hodder Gibson is grateful to Bright Red Publishing Ltd for collaborative work in preparation of this book and all SQA Past Paper, National 5 and Higher for CfE Model Paper titles 2014.

Typeset by PDQ Digital Media Solutions Ltd, Bungay, Suffolk NR35 1BY

Printed in the UK

A catalogue record for this title is available from the British Library

ISBN: 978-1-4718-3709-8

3 2

2015

Introduction

Study Skills – what you need to know to pass exams!

Pause for thought

Many students might skip quickly through a page like this. After all, we all know how to revise. Do you really though?

Think about this:

"IF YOU ALWAYS DO WHAT YOU ALWAYS DO, YOU WILL ALWAYS GET WHAT YOU HAVE ALWAYS GOT."

Do you like the grades you get? Do you want to do better? If you get full marks in your assessment, then that's great! Change nothing! This section is just to help you get that little bit better than you already are.

There are two main parts to the advice on offer here. The first part highlights fairly obvious things but which are also very important. The second part makes suggestions about revision that you might not have thought about but which WILL help you.

Part 1

DOH! It's so obvious but …

Start revising in good time

Don't leave it until the last minute – this will make you panic.

Make a revision timetable that sets out work time AND play time.

Sleep and eat!

Obvious really, and very helpful. Avoid arguments or stressful things too – even games that wind you up. You need to be fit, awake and focused!

Know your place!

Make sure you know exactly **WHEN and WHERE** your exams are.

Know your enemy!

Make sure you know what to expect in the exam.

How is the paper structured?

How much time is there for each question?

What types of question are involved?

Which topics seem to come up time and time again?

Which topics are your strongest and which are your weakest?

Are all topics compulsory or are there choices?

Learn by DOING!

There is no substitute for past papers and practice papers – they are simply essential! Tackling this collection of papers and answers is exactly the right thing to be doing as your exams approach.

Part 2

People learn in different ways. Some like low light, some bright. Some like early morning, some like evening / night. Some prefer warm, some prefer cold. But everyone uses their BRAIN and the brain works when it is active. Passive learning – sitting gazing at notes – is the most INEFFICIENT way to learn anything. Below you will find tips and ideas for making your revision more effective and maybe even more enjoyable. What follows gets your brain active, and active learning works!

Activity 1 – Stop and review

Step 1

When you have done no more than 5 minutes of revision reading STOP!

Step 2

Write a heading in your own words which sums up the topic you have been revising.

Step 3

Write a summary of what you have revised in no more than two sentences. Don't fool yourself by saying, "I know it, but I cannot put it into words". That just means you don't know it well enough. If you cannot write your summary, revise that section again, knowing that you must write a summary at the end of it. Many of you will have notebooks full of blue/black ink writing. Many of the pages will not be especially attractive or memorable so try to liven them up a bit with colour as you are reviewing and rewriting. **This is a great memory aid, and memory is the most important thing.**

Activity 2 — Use technology!

Why should everything be written down? Have you thought about "mental" maps, diagrams, cartoons and colour to help you learn? And rather than write down notes, why not record your revision material?

What about having a text message revision session with friends? Keep in touch with them to find out how and what they are revising and share ideas and questions.

Why not make a video diary where you tell the camera what you are doing, what you think you have learned and what you still have to do? No one has to see or hear it, but the process of having to organise your thoughts in a formal way to explain something is a very important learning practice.

Be sure to make use of electronic files. You could begin to summarise your class notes. Your typing might be slow, but it will get faster and the typed notes will be easier to read than the scribbles in your class notes. Try to add different fonts and colours to make your work stand out. You can easily Google relevant pictures, cartoons and diagrams which you can copy and paste to make your work more attractive and **MEMORABLE**.

Activity 3 – This is it. Do this and you will know lots!

Step 1

In this task you must be very honest with yourself! Find the SQA syllabus for your subject (www.sqa.org.uk). Look at how it is broken down into main topics called MANDATORY knowledge. That means stuff you MUST know.

Step 2

BEFORE you do ANY revision on this topic, write a list of everything that you already know about the subject. It might be quite a long list but you only need to write it once. It shows you all the information that is already in your long-term memory so you know what parts you do not need to revise!

Step 3

Pick a chapter or section from your book or revision notes. Choose a fairly large section or a whole chapter to get the most out of this activity.

With a buddy, use Skype, Facetime, Twitter or any other communication you have, to play the game "If this is the answer, what is the question?". For example, if you are revising Geography and the answer you provide is "meander", your buddy would have to make up a question like "What is the word that describes a feature of a river where it flows slowly and bends often from side to side?".

Make up 10 "answers" based on the content of the chapter or section you are using. Give this to your buddy to solve while you solve theirs.

Step 4

Construct a wordsearch of at least 10 X 10 squares. You can make it as big as you like but keep it realistic. Work together with a group of friends. Many apps allow you to make wordsearch puzzles online. The words and phrases can go in any direction and phrases can be split. Your puzzle must only contain facts linked to the topic you are revising. Your task is to find 10 bits of information to hide in your puzzle, but you must not repeat information that you used in Step 3. DO NOT show where the words are. Fill up empty squares with random letters. Remember to keep a note of where your answers are hidden but do not show your friends. When you have a complete puzzle, exchange it with a friend to solve each other's puzzle.

Step 5

Now make up 10 questions (not "answers" this time) based on the same chapter used in the previous two tasks. Again, you must find NEW information that you have not yet used. Now it's getting hard to find that new information! Again, give your questions to a friend to answer.

Step 6

As you have been doing the puzzles, your brain has been actively searching for new information. Now write a NEW LIST that contains only the new information you have discovered when doing the puzzles. Your new list is the one to look at repeatedly for short bursts over the next few days. Try to remember more and more of it without looking at it. After a few days, you should be able to add words from your second list to your first list as you increase the information in your long-term memory.

FINALLY! Be inspired...

Make a list of different revision ideas and beside each one write **THINGS I HAVE** tried, **THINGS I WILL** try and **THINGS I MIGHT** try. Don't be scared of trying something new.

And remember – "FAIL TO PREPARE AND PREPARE TO FAIL!"

National 5 Mathematics

The course

The National 5 Mathematics course aims to enable you to develop the ability to:

- select and apply mathematical techniques in a variety of mathematical and real-life situations
- manipulate abstract terms in order to solve problems and to generalise
- interpret, communicate and manage information in mathematical form
- use mathematical language and explore mathematical ideas.

Before starting this course you should already have the knowledge, understanding and skills required to achieve a good pass in National 4 Mathematics and/or be proficient in equivalent experiences and outcomes. This course enables you to further develop your knowledge, understanding and skills in algebra, geometry, trigonometry, numeracy, statistics and reasoning. The course content is summarised below.

<table>
<tr><th>Algebra</th><th>Geometry</th><th>Trigonometry</th></tr>
<tr><td rowspan="3">• Expanding brackets
• Factorising
• Completing the square
• Algebraic fractions
• Equation of straight line
• Equations and inequations
• Simultaneous equations
• Change of subject of formulae
• Graphs of quadratic functions
• Quadratic equations</td><td>• Gradient
• Arc and sector of circle
• Volume (including significant figures)
• Pythagoras' theorem
• Properties of shapes
• Similarity
• Vectors</td><td>• Graphs
• Equations
• Identities
• Area of triangle, sine rule, cosine rule, bearings</td></tr>
<tr><th>Numeracy</th><th>Statistics</th></tr>
<tr><td>• Surds
• Indices
• Percentages
• Fractions</td><td>• Interquartile range, standard deviation
• Scattergraphs; equation of line of best fit</td></tr>
<tr><th colspan="3">Reasoning</th></tr>
<tr><td colspan="3">• Interpreting a situation where mathematics can be used and identifying a strategy.
• Explaining a solution and/or relating it to context.</td></tr>
</table>

Assessment

To gain the Course award, you must pass the three Units – Expressions & Formulae, Relationships and Applications – as well as the examination. The Units are assessed internally on a pass/fail basis and the examination is set and marked externally by the SQA. It tests skills beyond the minimum competence required for the Units.

The number of marks and the times allotted for the examination papers are as follows:

Paper 1 (non-calculator)	40 marks	1 hour
Paper 2	50 marks	1 hour 30 minutes

The Course award is graded A-D, the grade being determined by the total mark you score in the examination.

Some tips for achieving a good mark

- **DOING** maths questions is the most effective use of your study time. You will benefit much more from spending 30 minutes doing maths questions than spending several hours copying out notes or reading a maths textbook.
- Practise doing the type of questions that are likely to appear in the exam. Work through these practice papers and similar questions from past Credit Level and Intermediate 2 papers. Use the marking instructions to check your answers and to understand what the examiners are looking for. Ask your teacher for help if you get stuck.
- **SHOW ALL WORKING CLEARLY.** The instructions on the front of the exam paper state that "Full credit will only be given where the solution contains appropriate working". A "correct" answer with no working may only be awarded partial marks or even no marks at all. An incomplete answer will be awarded marks for any appropriate working. Attempt every question, even if you are not sure whether you are correct or not. Your solution may contain working which will gain some marks. A blank response is certain to be awarded no marks. Never score out working unless you have something better to replace it with.
- Communication is very important in presenting solutions to questions. Diagrams are often a good way of conveying information and enabling markers to understand your working. Where a diagram is included in a question, it is often good practice to copy it and show the results of your working on the copy.

- In Paper 1, you have to carry out calculations without a calculator. Candidates' performance in number skills is often disappointing, and costs many of them valuable marks. Ensure that you practise your number skills regularly, especially within questions testing Course content.
- In Paper 2, you will be allowed to use a calculator. Always use **your own** calculator. Different calculators often function in slightly different ways, so make sure that you know how to operate yours. Having to use a calculator that you are unfamiliar with on the day of the exam will disadvantage you.
- Prepare thoroughly to tackle questions from **all** parts of the course. Numerical and algebraic fractions, graphs of quadratic fractions, surds, indices and trigonometric identities are topics that often cause candidates problems. Be prepared to put extra effort into mastering these topics.

Some common errors to avoid

	Common error	**Correct answer**
Converse of Pythagoras' Theorem e.g. Prove that triangle ABC is right angled. 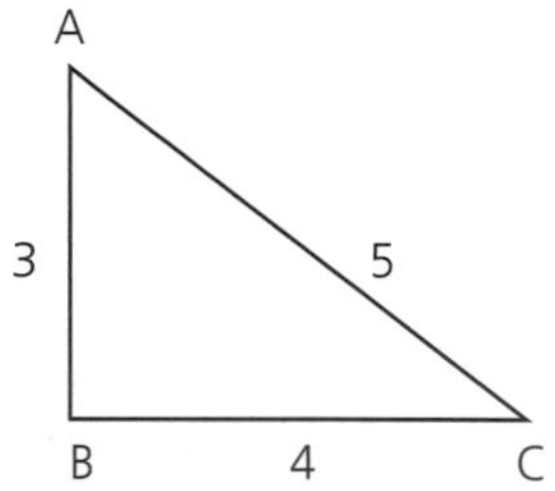	Don't start by assuming what you are trying to prove is true. $AC^2 = AB^2 + BC^2$ $AC^2 = 3^2 + 4^2 = 9 + 16 = 25$ $AC = \sqrt{25} = 5$ so triangle ABC is right angled by the Converse of Pythagoras' Theorem.	Don't state that $AC^2 = AB^2 + BC^2$ until you have the evidence to prove that it is true. $AC^2 = 5^2 = 25$ $AB^2 + BC^2 = 3^2 + 4^2 = 9 + 16 = 25$ so $AC^2 = AB^2 + BC^2$ so triangle ABC is right angled by the Converse of Pythagoras' Theorem.
Similarity (area and volume) e.g. 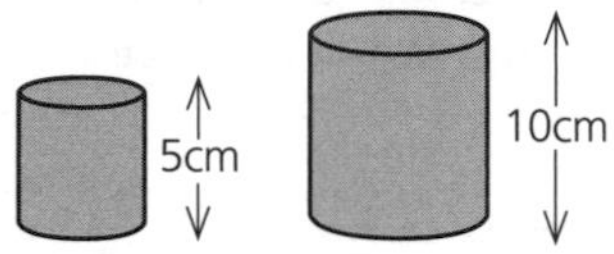 Theses cylinders are mathematically similar. The volume of the small one is $60cm^3$. Calculate the volume of the large one.	Don't use the linear scale factor to calculate the volume (or area) of a similar shape. Scale factor = 2 Volume = 2 × 60 = $120cm^3$	Remember that volume factor = $(\text{linear factor})^3$ area factor = $(\text{linear factor})^2$ Scale factor = 2 Volume = $2^3 \times 60 = 480cm^3$
Reverse use of percentage e.g. After a 5% pay rise, Ann now earns £252 per week. Calculate her weekly pay before the rise.	Increase = 5% of old pay **NOT** 5% of new pay Increase = 5% of £252 = £12·60 Old pay = £252 - £12·60 = £239·40	New pay = (100% + 5%) of old pay New pay =105% of old pay = £252 1% of old pay = £252 ÷ 105 = £2·40 Old pay = 100%= £2·40×100 =£240
Interpreting statistics e.g. Jack and Jill sat tests in the same eight subjects. Jack's mean mark was 76 and his standard deviation was 13. Jill's mean mark was 59 and her standard deviation was 21. Make two valid comments comparing the performance of Jack and Jill in the tests.	This answer does not show that you **understand** the meaning of mean and standard deviation. Jack has a higher mean mark but a lower standard deviation than Jill.	Your interpretation of the figures must show that you **understand** that mean is an average and that standard deviation is a measure of spread. On average Jack performed better than Jill as his mean mark was higher. Jack's performance was more consistent than Jill's as the standard deviation of his marks was lower.

Good luck!

Remember that the rewards for passing National 5 Mathematics are well worth it! Your pass will help you get the future you want for yourself. In the exam, be confident in your own ability, if you're not sure how to answer a question trust your instincts and just give it a go anyway – keep calm and don't panic! GOOD LUCK!

NATIONAL 5

2013 Specimen Question Paper

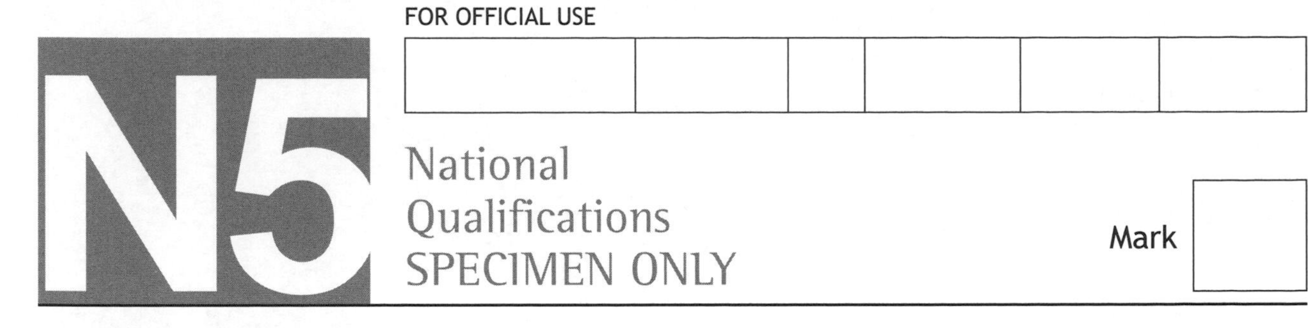

SQ29/N5/01

Mathematics Paper 1 (Non-Calculator)

Date — Not applicable

Duration — 1 hour

Fill in these boxes and read what is printed below.

Full name of centre

Town

Forename(s)

Surname

Number of seat

Date of birth

Day Month Year

D D M M Y Y

Scottish candidate number

Total marks — 40

You may NOT use a calculator.

Attempt ALL questions.

Use **blue** or **black** ink. Pencil may be used for graphs and diagrams only.

Write your working and answers in the spaces provided. Additional space for answers is provided at the end of this booklet. If you use this space, write clearly the number of the question you are attempting.

Square-ruled paper is provided at the back of this booklet.

Full credit will be given only to solutions which contain appropriate working.

State the units for your answer where appropriate.

Before leaving the examination room you must give this booklet to the Invigilator. If you do not, you may lose all the marks for this paper.

FORMULAE LIST

The roots of $ax^2 + bx + c = 0$ are $x = \dfrac{-b \pm \sqrt{(b^2 - 4ac)}}{2a}$

Sine rule: $\dfrac{a}{\sin A} = \dfrac{b}{\sin B} = \dfrac{c}{\sin C}$

Cosine rule: $a^2 = b^2 + c^2 - 2bc\cos A$ or $\cos A = \dfrac{b^2 + c^2 - a^2}{2bc}$

Area of a triangle: $A = \frac{1}{2}ab\sin C$

Volume of a sphere: $V = \frac{4}{3}\pi r^3$

Volume of a cone: $V = \frac{1}{3}\pi r^2 h$

Volume of a pyramid: $V = \frac{1}{3}Ah$

Standard deviation: $s = \sqrt{\dfrac{\Sigma(x-\overline{x})^2}{n-1}} = \sqrt{\dfrac{\Sigma x^2 - (\Sigma x)^2/n}{n-1}}$, where n is the sample size.

MARKS | DO NOT WRITE IN THIS MARGIN

1. Evaluate

$$2\frac{3}{8} \div \frac{5}{16}.$$ 2

$= \frac{19}{8} \div \frac{5}{16}$

2. Multiply out the brackets and collect like terms

$$(2x+3)(x^2-4x+1).$$ 3

$= 2x(x^2-4x+1) + 3(x^2-4x+1)$

$= 2x^3 - 8x^2 + 2x + 3x^2 - 12x + 3$

$= 2x^3 - 5x^2 - 10x + 3$

3. Two forces acting on a rocket are represented by vectors **u** and **v**.

$$\mathbf{u} = \begin{pmatrix} 2 \\ -5 \\ -3 \end{pmatrix} \text{ and } \mathbf{v} = \begin{pmatrix} 7 \\ 4 \\ -1 \end{pmatrix}.$$

Calculate $|\mathbf{u} + \mathbf{v}|$, the magnitude of the resultant force.

Express your answer as a surd in its simplest form. 3

$|u + v|$

$\begin{pmatrix} 2 \\ -5 \\ -3 \end{pmatrix} + \begin{pmatrix} 7 \\ 4 \\ -1 \end{pmatrix} = \begin{pmatrix} 9 \\ -1 \\ -4 \end{pmatrix}$

$= \sqrt{9^2 + (-1)^2 + (-4)^2}$

$= \sqrt{81 + 1 + 16}$

$= \sqrt{98}$

$= 7\sqrt{2}$

MARKS | DO NOT WRITE IN THIS MARGIN

4. Solve the equation

$$2x^2 + 7x - 15 = 0.$$

$ax^2 + bx + c = 0$ $\quad x = \dfrac{-b \pm \sqrt{(b^2 - 4ac}}{2a}$

3

$= \dfrac{-7 \pm \sqrt{(7^2 - 4 \times 2 \times (-15)}}{2 \times 2}$ $\quad = \dfrac{-7 \pm \sqrt{(49 - (-120)}}{}$

$=$

5. Express $\dfrac{4}{\sqrt{6}}$ with a rational denominator in its simplest form.

2

$\dfrac{4}{\sqrt{6}} \times \dfrac{\sqrt{6}}{\sqrt{6}} = \dfrac{4\sqrt{6}}{6}$

MARKS | DO NOT WRITE IN THIS MARGIN

6. Teams in a quiz answer questions on film and sport.

This scattergraph shows the scores of some of the teams.

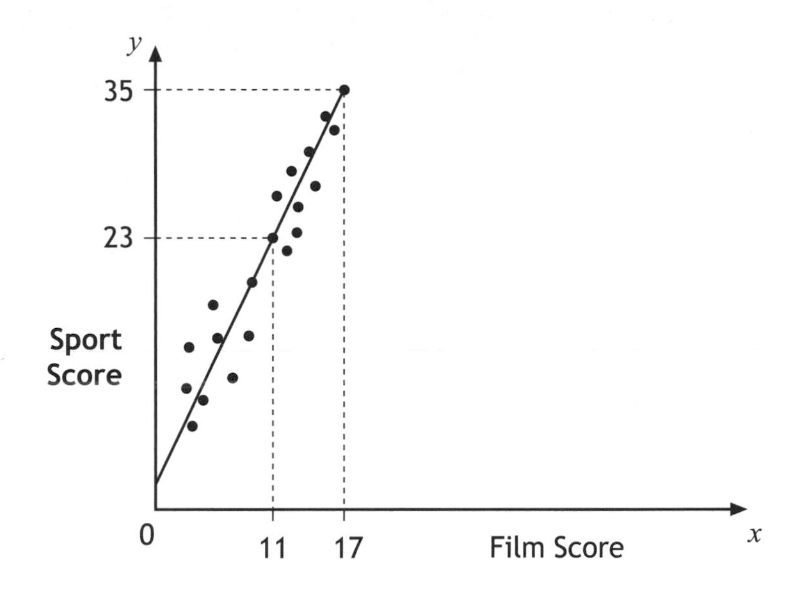

A line of best fit is drawn as shown.

(a) Find the equation of this straight line. **3**

$y = 2x + 1$

(b) Use this equation to estimate the sports score for a team with a film score of 8. **1**

Total marks 4

MARKS | DO NOT WRITE IN THIS MARGIN

7. (a) Multiply out the brackets and simplify:

$$x^{\frac{1}{2}}\left(x^{-\frac{3}{2}}+x^{-\frac{1}{2}}\right).$$

2

(b) Find the exact value of this expression when $x=6$. 1

Total marks 3

8. Change the subject of the formula $p=\frac{mv^2}{2}$ to v. 3

$p = \frac{mv^2}{2}$

$\frac{mv^2}{2} = p$

$mv^2 = 2p$

$v^2 = \frac{2p}{m}$

$v = \sqrt{\frac{2p}{m}}$

MARKS | DO NOT WRITE IN THIS MARGIN

9. A parabola has equation $y = x^2 - 8x + 19$.

(a) Write the equation in the form $y = (x - p)^2 + q$. **2**

(b) Sketch the graph of $y = x^2 - 8x + 19$, showing the coordinates of the turning point and the point of intersection with the y-axis. **3**

Total marks 5

MARKS | DO NOT WRITE IN THIS MARGIN

10. Brian and Bob visit a ski resort. Brian buys 3 full passes and 4 restricted passes. The total cost of his passes is £185.

(a) Write down an equation to illustrate this information. **1**

(b) Bob buys 2 full passes and 3 restricted passes.

The total cost of his passes is £130.

Write down an equation to illustrate this information. **1**

(c) Find the cost of a restricted pass and the cost of a full pass. **3**

Total marks 5

11. Express

$$\frac{4}{x+2} - \frac{3}{x-4}, \qquad x \neq -2,\ x \neq 4$$

as a single fraction in its simplest form. **3**

MARKS DO NOT WRITE IN THIS MARGIN

12. A cylindrical pipe has water in it as shown.

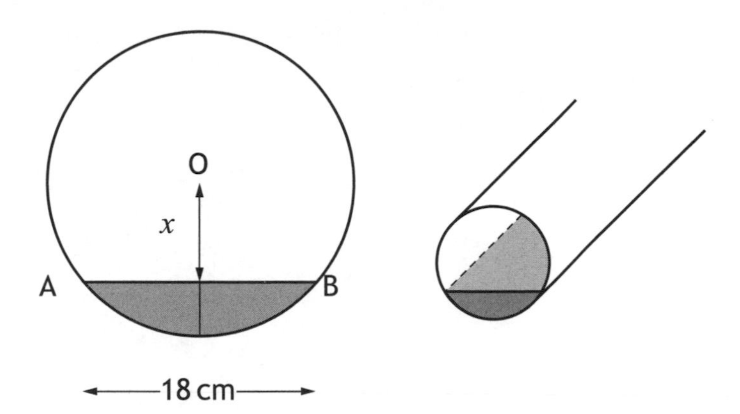

The depth of the water at the deepest point is 5 centimetres.

The width of the water surface, AB, is 18 centimetres.

The radius of the pipe is r centimetres.

The distance from the centre, O, of the pipe to the water surface is x centimetres.

(a) Write down an expression for x in terms of r. **1**

(b) Calculate r, the radius of the pipe. **3**

Total marks 4

[END OF SPECIMEN QUESTION PAPER]

ADDITIONAL SPACE FOR ANSWERS

ADDITIONAL SPACE FOR ANSWERS

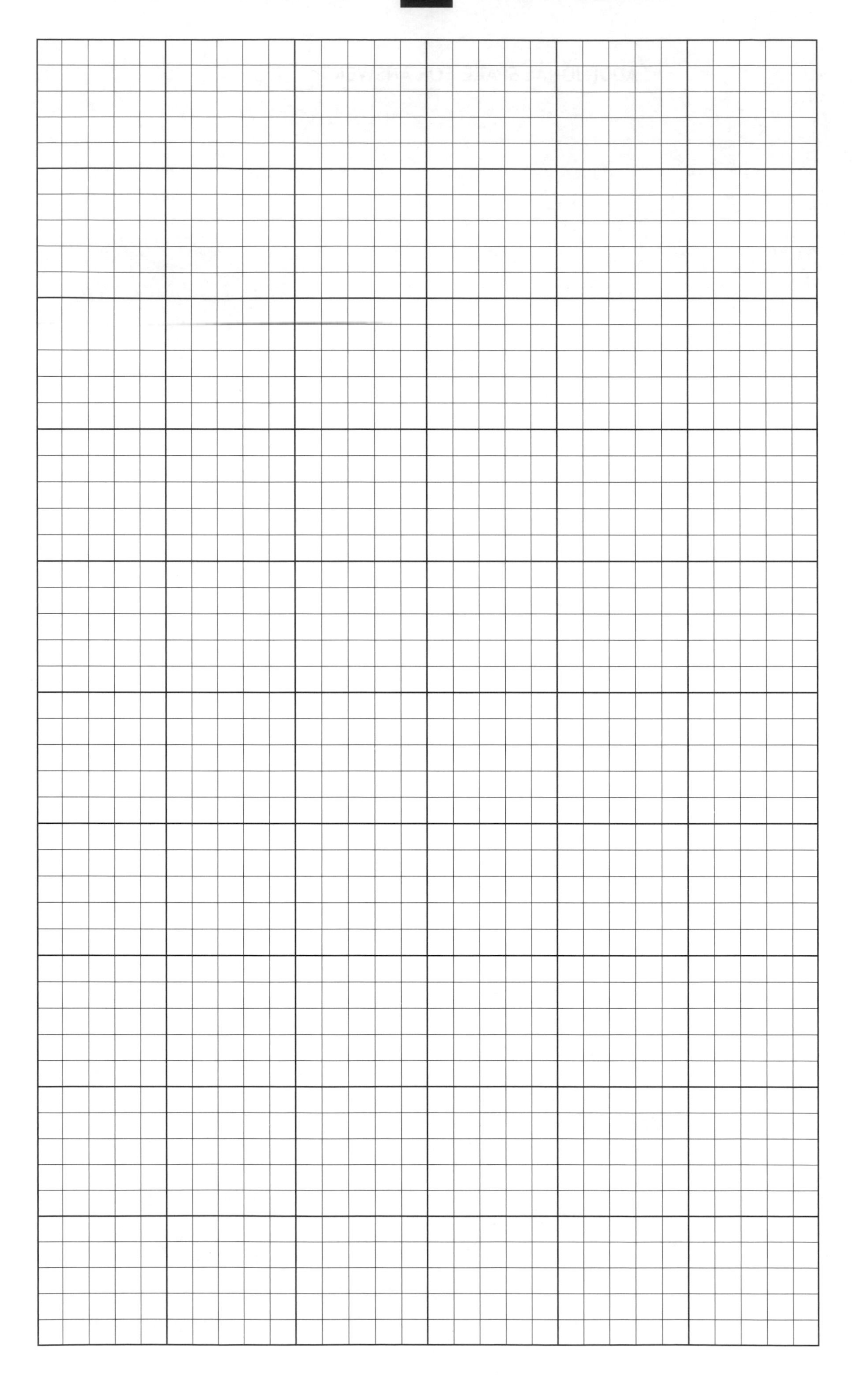

FOR OFFICIAL USE

N5

National
Qualifications
SPECIMEN ONLY

Mark

SQ29/N5/02

Mathematics
Paper 2

Date — Not applicable

Duration — 1 hour and 30 minutes

Fill in these boxes and read what is printed below.

Full name of centre

Town

Forename(s)

Surname

Number of seat

Date of birth

Day Month Year

D D M M Y Y

Scottish candidate number

Total marks — 50

You may use a calculator.

Attempt ALL questions.

Use **blue** or **black** ink. Pencil may be used for graphs and diagrams only.

Write your working and answers in the spaces provided. Additional space for answers is provided at the end of this booklet. If you use this space, write clearly the number of the question you are attempting.

Square-ruled paper is provided at the back of this booklet.

Full credit will be given only to solutions which contain appropriate working.

State the units for your answer where appropriate.

Before leaving the examination room you must give this booklet to the Invigilator.
If you do not, you may lose all the marks for this paper.

FORMULAE LIST

The roots of $ax^2 + bx + c = 0$ are $x = \dfrac{-b \pm \sqrt{(b^2 - 4ac)}}{2a}$

Sine rule: $\dfrac{a}{\sin A} = \dfrac{b}{\sin B} = \dfrac{c}{\sin C}$

Cosine rule: $a^2 = b^2 + c^2 - 2bc\cos A$ or $\cos A = \dfrac{b^2 + c^2 - a^2}{2bc}$

Area of a triangle: $A = \frac{1}{2}ab\sin C$

Volume of a sphere: $V = \frac{4}{3}\pi r^3$

Volume of a cone: $V = \frac{1}{3}\pi r^2 h$

Volume of a pyramid: $V = \frac{1}{3}Ah$

Standard deviation: $s = \sqrt{\dfrac{\Sigma(x - \overline{x})^2}{n - 1}} = \sqrt{\dfrac{\Sigma x^2 - (\Sigma x)^2/n}{n - 1}}$, where n is the sample size.

MARKS DO NOT WRITE IN THIS MARGIN

1. Beth normally cycles a total distance of 56 miles per week.

   She increases her distance by 15% each week for the next three weeks.

   How many miles will she cycle in the third week? **3**

2. There are 3×10^5 platelets per millilitre of blood.

   On average, a person has 5·5 litres of blood.

   On average, how many platelets does a person have in their blood?

   Give your answer in scientific notation. **2**

MARKS | DO NOT WRITE IN THIS MARGIN

3. In the diagram, OABCDE is a regular hexagon with centre M.

Vectors $\mathbf{a}$ and $\mathbf{b}$ are represented by $\overrightarrow{OA}$ and $\overrightarrow{OB}$ respectively.

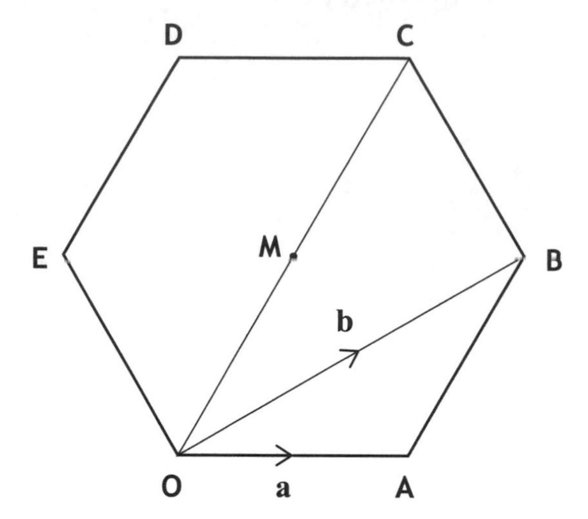

(a) Express $\overrightarrow{AB}$ in terms of $\mathbf{a}$ and $\mathbf{b}$. **1**

(b) Express $\overrightarrow{OC}$ in terms of $\mathbf{a}$ and $\mathbf{b}$. **1**

Total marks 2

MARKS | DO NOT WRITE IN THIS MARGIN

4. The graph with equation $y = kx^2$ is shown below.

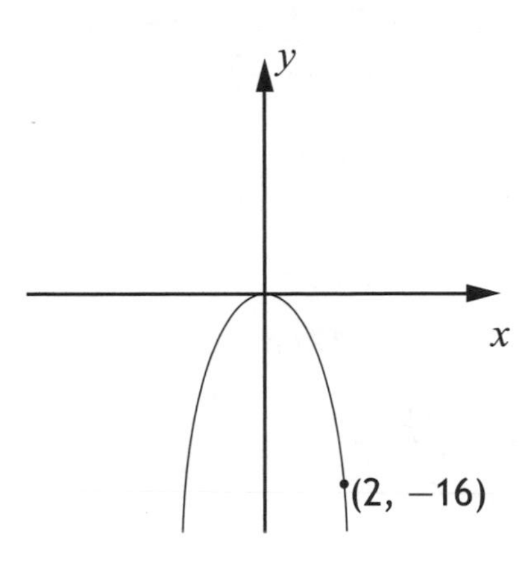

The point (2, −16) lies on the graph.

Determine the value of k. **2**

5. In triangle PQR, PQ = 8 centimetres, QR = 3 centimetres and angle PQR = 120°.

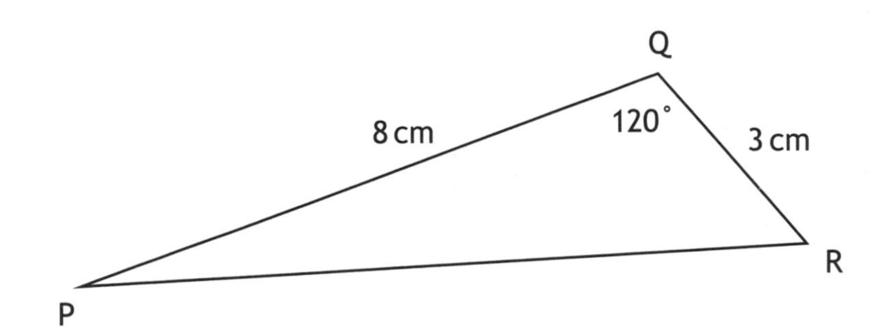

Calculate the length of PR. **3**

MARKS DO NOT WRITE IN THIS MARGIN

6. A child's toy is in the shape of a hemisphere with a cone on top, as shown in the diagram.

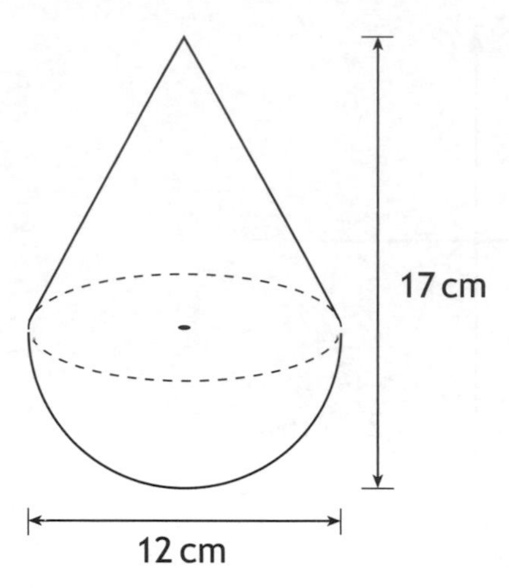

The toy is 12 centimetres wide and 17 centimetres high.

Calculate the volume of the toy.

Give your answer correct to 2 significant figures. **5**

MARKS | DO NOT WRITE IN THIS MARGIN

7. This year Adèle paid £465 for her car insurance.

This is an increase of 20% on last year's payment.

How much did Adèle pay last year? **3**

8. A frozen food company uses machines to pack sprouts into bags.

A sample of six bags is taken from Machine A and the number of sprouts in each bag is counted.

The results are shown below.

23 19 21 20 19 24

(a) Calculate the mean and standard deviation of this sample. **3**

(b) Another sample of six bags is taken from Machine B.

This sample has a mean of 19 and a standard deviation of 2·3.

Write down two valid comparisons between the samples. **2**

Total marks 5

MARKS | DO NOT WRITE IN THIS MARGIN

9. Screenwash is available in two different sized bottles, 'Mini' and 'Maxi'.

The bottles are mathematically similar.

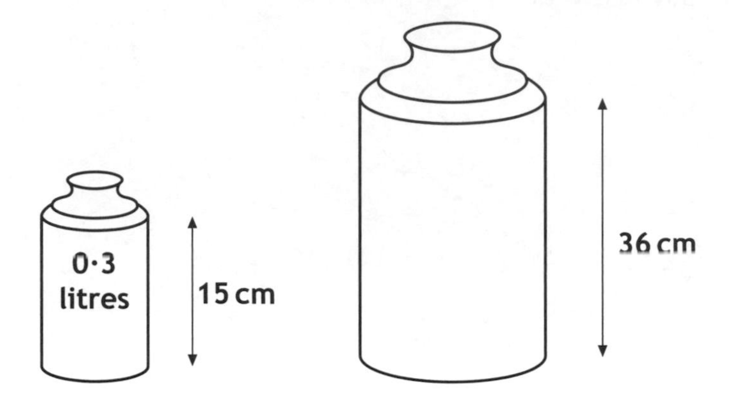

Calculate the volume of the 'Maxi' bottle. **3**

MARKS | DO NOT WRITE IN THIS MARGIN

10. Part of the graph of $y = a\cos x^\circ + b$ is shown below.

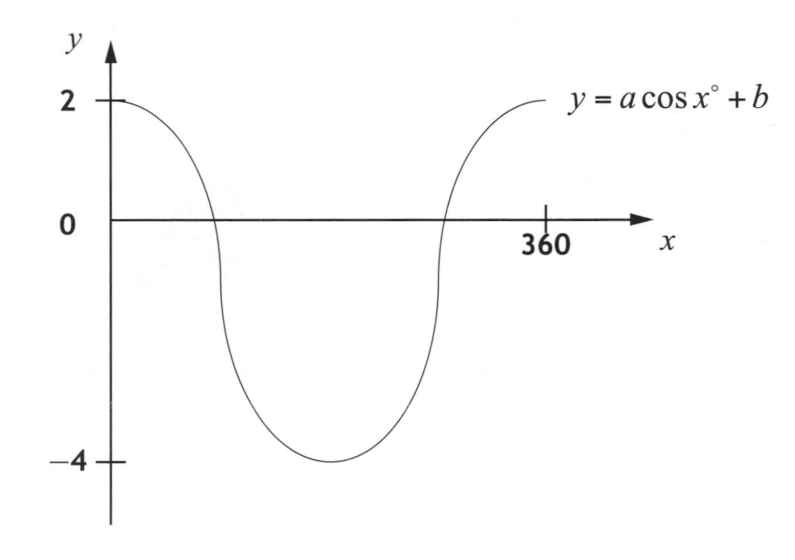

(a) Explain how you can tell from the graph that $a = 3$ and $b = -1$. **2**

(b) Calculate the x-coordinates of the points where the graph cuts the x-axis. **4**

Total marks 6

MARKS DO NOT WRITE IN THIS MARGIN

11. A cone is formed from a paper circle with a sector removed as shown.

The radius of the paper circle is 40 centimetres.

Angle AOB is 110°.

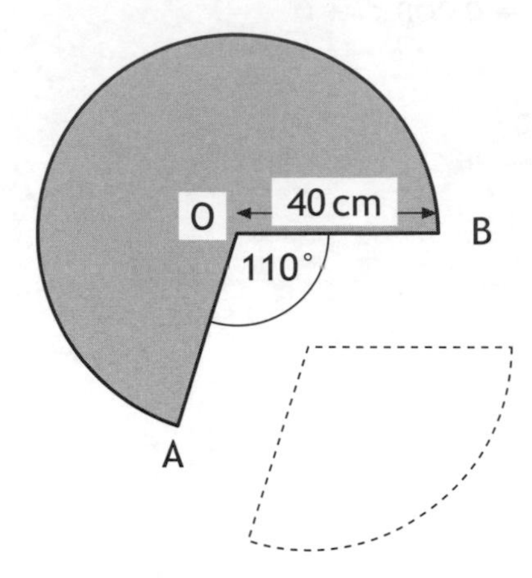

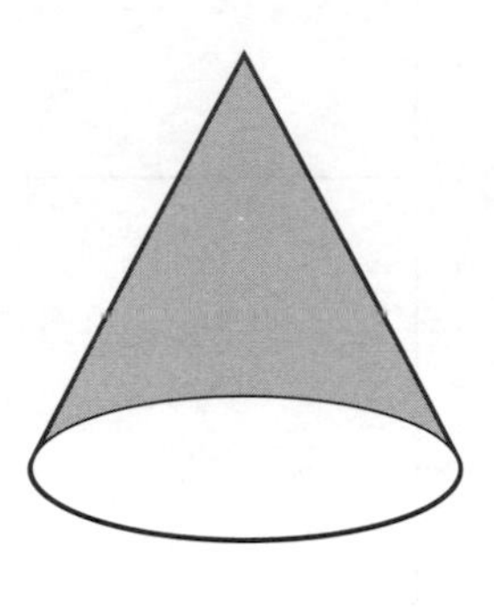

(a) Calculate the area of the sector removed from the circle. **3**

(b) Calculate the circumference of the base of the cone. **3**

Total marks 6

MARKS DO NOT WRITE IN THIS MARGIN

12. Find the range of values of p such that the equation $px^2 - 2x + 3 = 0,\ p \neq 0$, has no real roots. **4**

MARKS | DO NOT WRITE IN THIS MARGIN

13. A yacht sails from a harbour H to a point C, then to a point D as shown below.

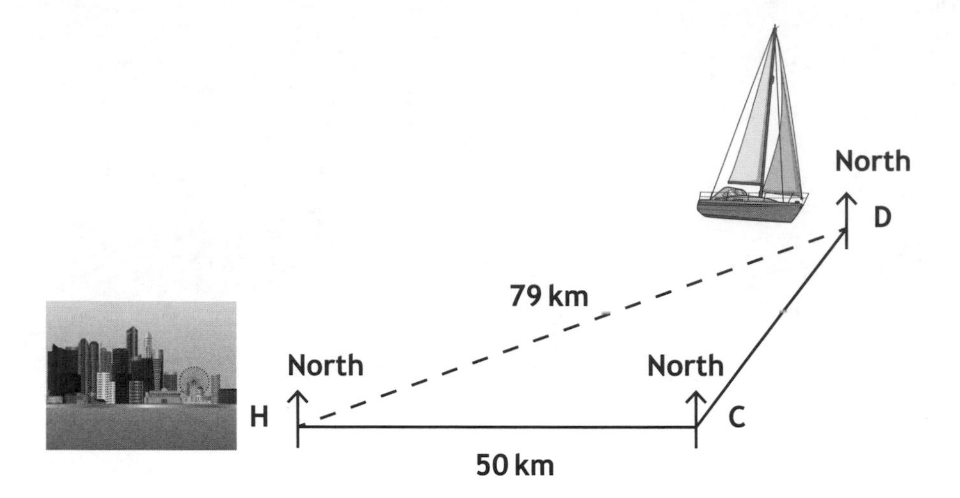

C is 50 kilometres due east of H.

D is on a bearing of 040° from C and is 79 kilometres from H.

(a) Calculate the size of angle CDH. **4**

(b) Hence, calculate the bearing on which the yacht must sail to return directly to the harbour. **2**

Total marks 6

[END OF SPECIMEN QUESTION PAPER]

ADDITIONAL SPACE FOR ANSWERS

ADDITIONAL SPACE FOR ANSWERS

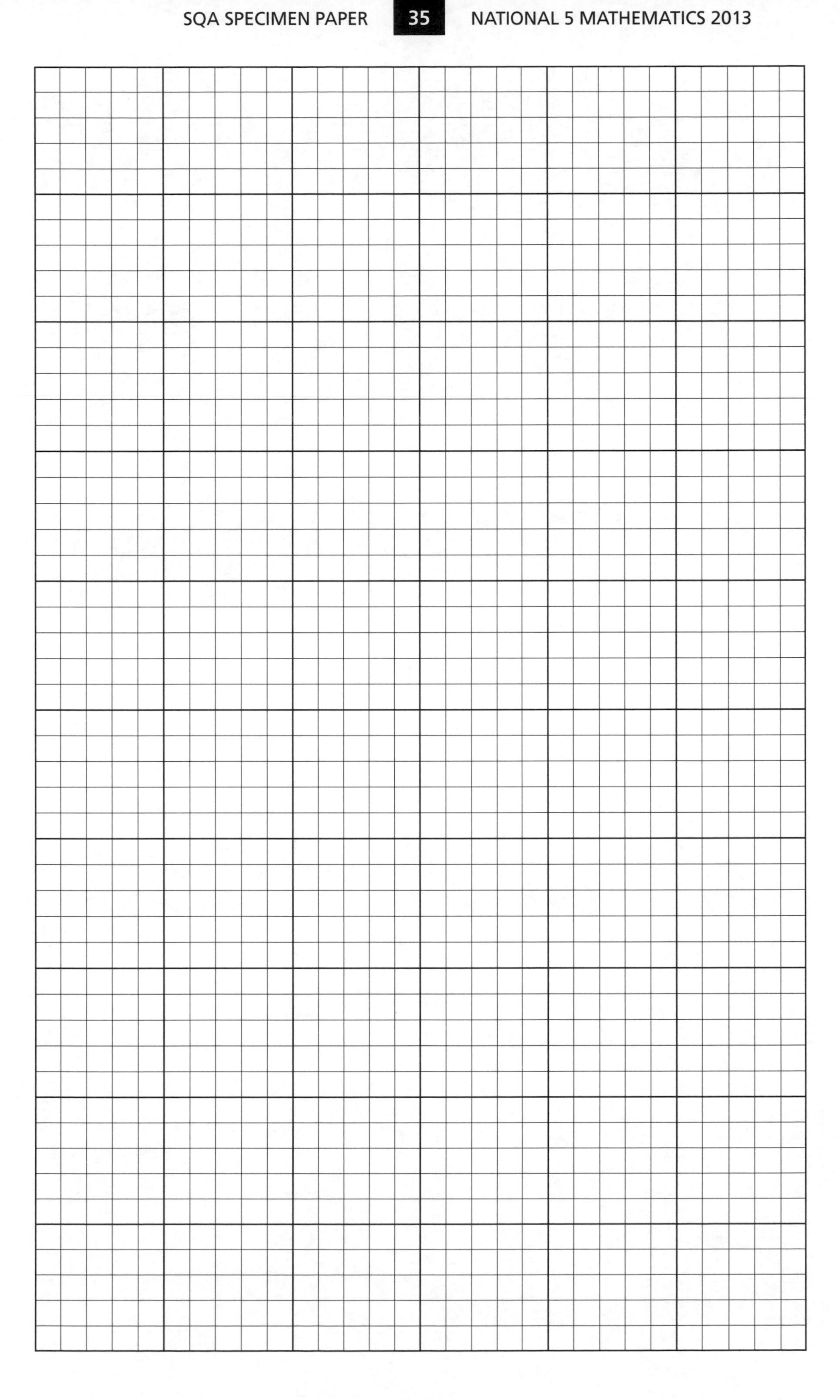

NATIONAL 5

Model Paper 1

Whilst this Model Practice Paper has been specially commissioned by Hodder Gibson for use as practice for the National 5 exams, the key reference documents remain the SQA Specimen Paper 2013 and the SQA Past Paper 2014.

Mathematics
Paper 1
(Non-Calculator)

Duration — 1 hour

Total marks — 40

You may NOT use a calculator.

Attempt ALL questions.

Use **blue** or **black** ink. Pencil may be used for graphs and diagrams only.

Write your working and answers in the spaces provided. Additional space for answers is provided at the end of this booklet. If you use this space, write clearly the number of the question you are attempting.

Square-ruled paper is provided at the back of this booklet.

Full credit will be given only to solutions which contain appropriate working.

State the units for your answer where appropriate.

Before leaving the examination room you must give this booklet to the Invigilator. If you do not, you may lose all the marks for this paper.

FORMULAE LIST

The roots of $ax^2 + bx + c = 0$ are $x = \dfrac{-b \pm \sqrt{(b^2 - 4ac)}}{2a}$

Sine rule: $\dfrac{a}{\sin A} = \dfrac{b}{\sin B} = \dfrac{c}{\sin C}$

Cosine rule: $a^2 = b^2 + c^2 - 2bc\cos A$ or $\cos A = \dfrac{b^2 + c^2 - a^2}{2bc}$

Area of a triangle: $A = \frac{1}{2}ab\sin C$

Volume of a sphere: $V = \frac{4}{3}\pi r^3$

Volume of a cone: $V = \frac{1}{3}\pi r^2 h$

Volume of a pyramid: $V = \frac{1}{3}Ah$

Standard deviation: $s = \sqrt{\dfrac{\Sigma(x - \bar{x})^2}{n-1}} = \sqrt{\dfrac{\Sigma x^2 - (\Sigma x)^2/n}{n-1}}$, where n is the sample size.

MARKS DO NOT WRITE IN THIS MARGIN

1. Evaluate

$$4\frac{1}{3}-1\frac{1}{2}.$$ 2

2. Expand and simplify

$$(3x-2)(2x^2+x+5).$$ 3

3. Change the subject of the formula to *m*.

$$L=\frac{\sqrt{m}}{k}$$ 2

MARKS DO NOT WRITE IN THIS MARGIN

4. The diagram shows a tiling of congruent triangles.

Vectors **u** and **v** are represented by $\overrightarrow{AB}$ and $\overrightarrow{AF}$ respectively.

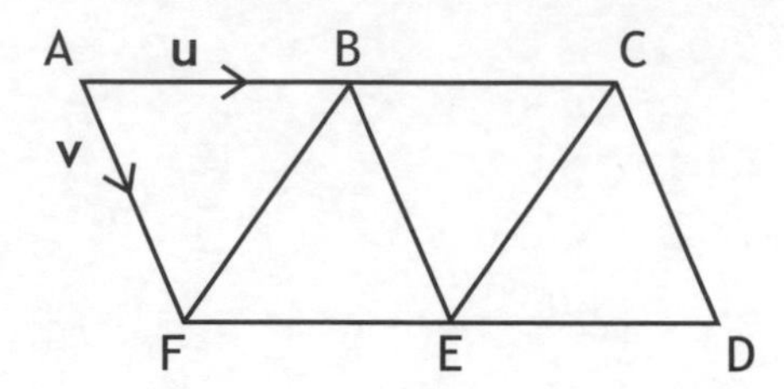

(a) Express $\overrightarrow{AD}$ in terms of **u** and **v**.

(b) Express $\overrightarrow{CE}$ in terms of **u** and **v**. **1**

Total marks 2

5.

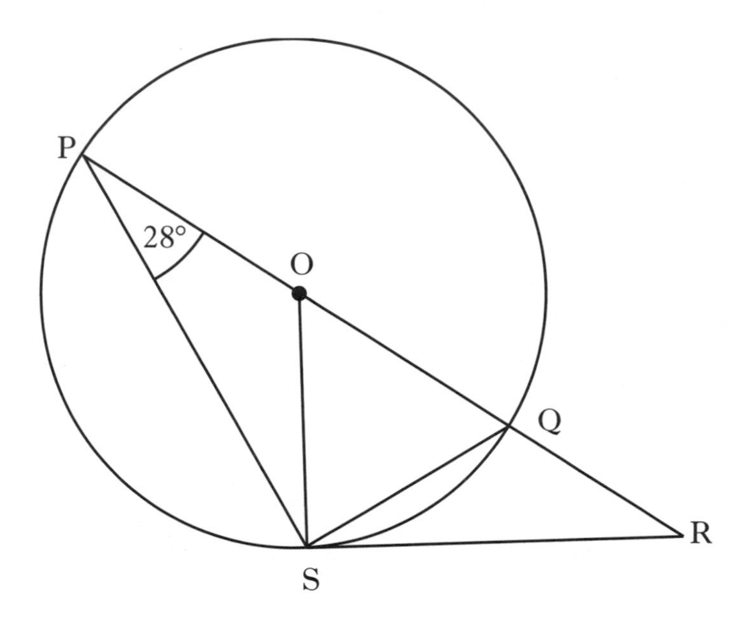

In the above diagram,

- O is the centre of the circle
- PQ is a diameter of the circle
- PQR is a straight line
- RS is a tangent to the circle at S
- angle QPS is 28°.

Calculate the size of angle QRS. **3**

MARKS | DO NOT WRITE IN THIS MARGIN

6. Express $\dfrac{3y^2-6y}{y^2+y-6}$ in its simplest form. **3**

7. Evaluate $9^{\frac{3}{2}}$. **2**

8. The diagram shows part of the graph of $y=5+4x-x^2$.

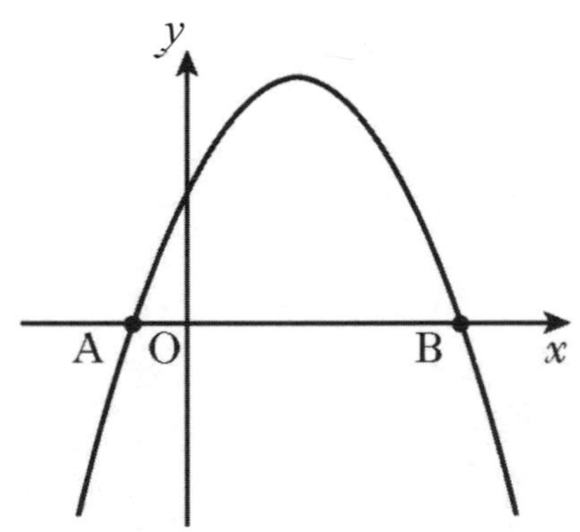

$y=5+4x-x^2$

A is the point (-1, 0).

B is the point (5, 0).

(a) State the equation of the axis of symmetry of the graph. **2**

(b) Hence, find the maximum value of $y=5+4x-x^2$. **2**

Total marks 4

MARKS | DO NOT WRITE IN THIS MARGIN

9. The graph below shows two straight lines.

- $y = 2x - 3$
- $x + 2y = 14$

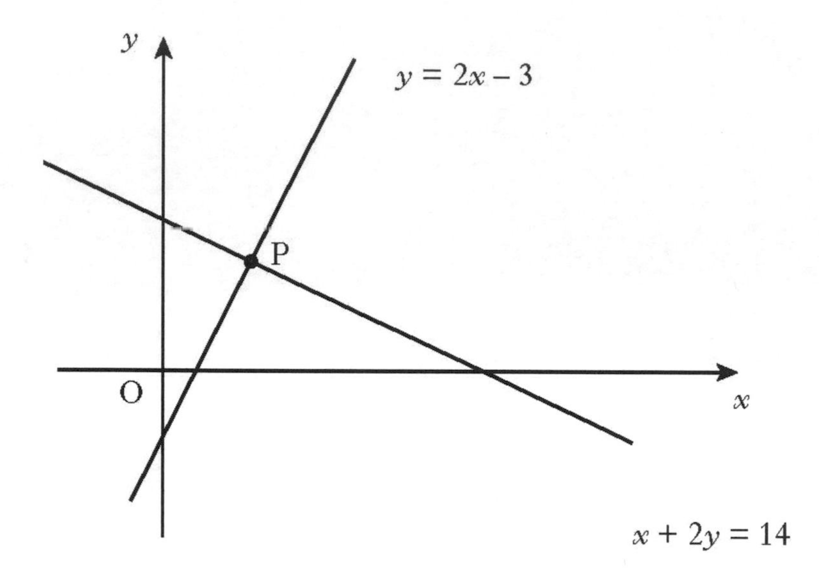

The lines intersect at the point P.

Find, **algebraically**, the coordinates of P. **4**

MARKS DO NOT WRITE IN THIS MARGIN

10. Part of the graph of $y = a\cos bx^o$ is shown in the diagram.

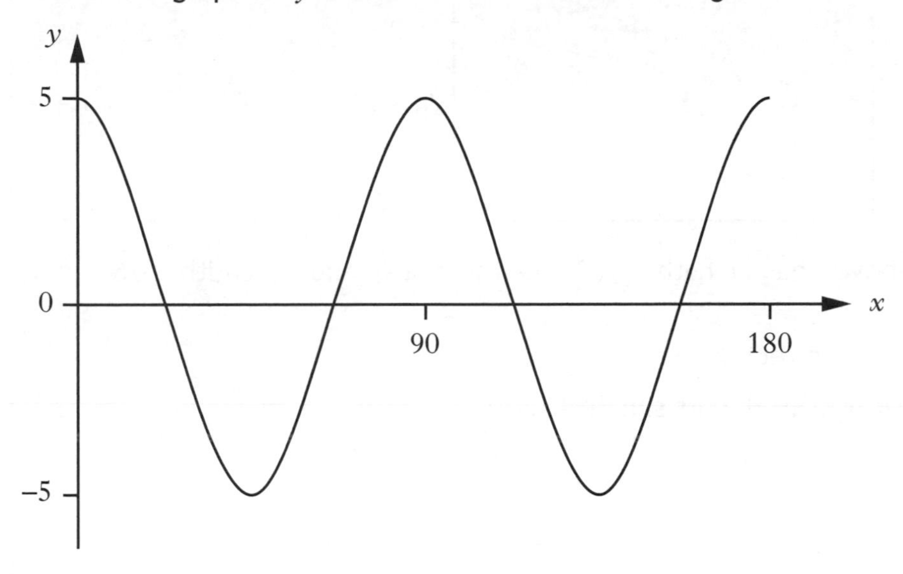

State the values of a and b. **2**

11.

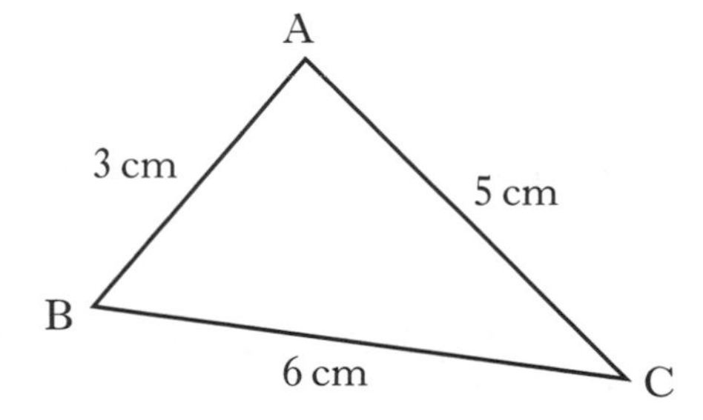

In triangle ABC, show that cos B = $\frac{5}{9}$. **3**

MARKS DO NOT WRITE IN THIS MARGIN

12.

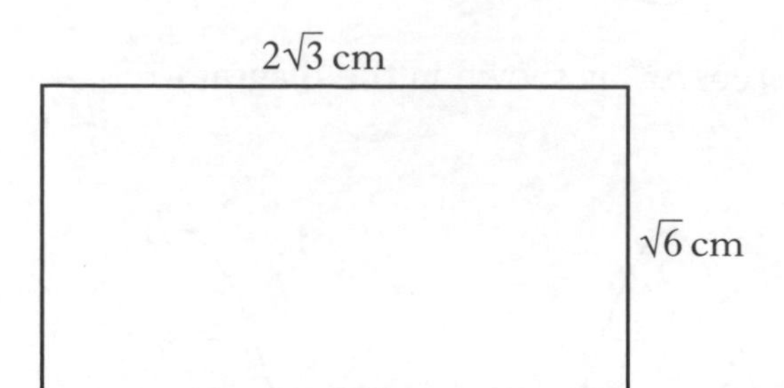

The rectangle above has length $2\sqrt{3}$ centimetres and breadth $\sqrt{6}$ centimetres.

Calculate the area of the rectangle.

Express your answer as a surd in its simplest form. **3**

13. Simplify $\frac{3}{m} + \frac{4}{m+1}$. **3**

14. Prove that the roots of the equation $2x^2 + 8x + 5 = 0$ are real and irrational. **4**

[END OF MODEL PRACTICE PAPER]

ADDITIONAL SPACE FOR ANSWERS

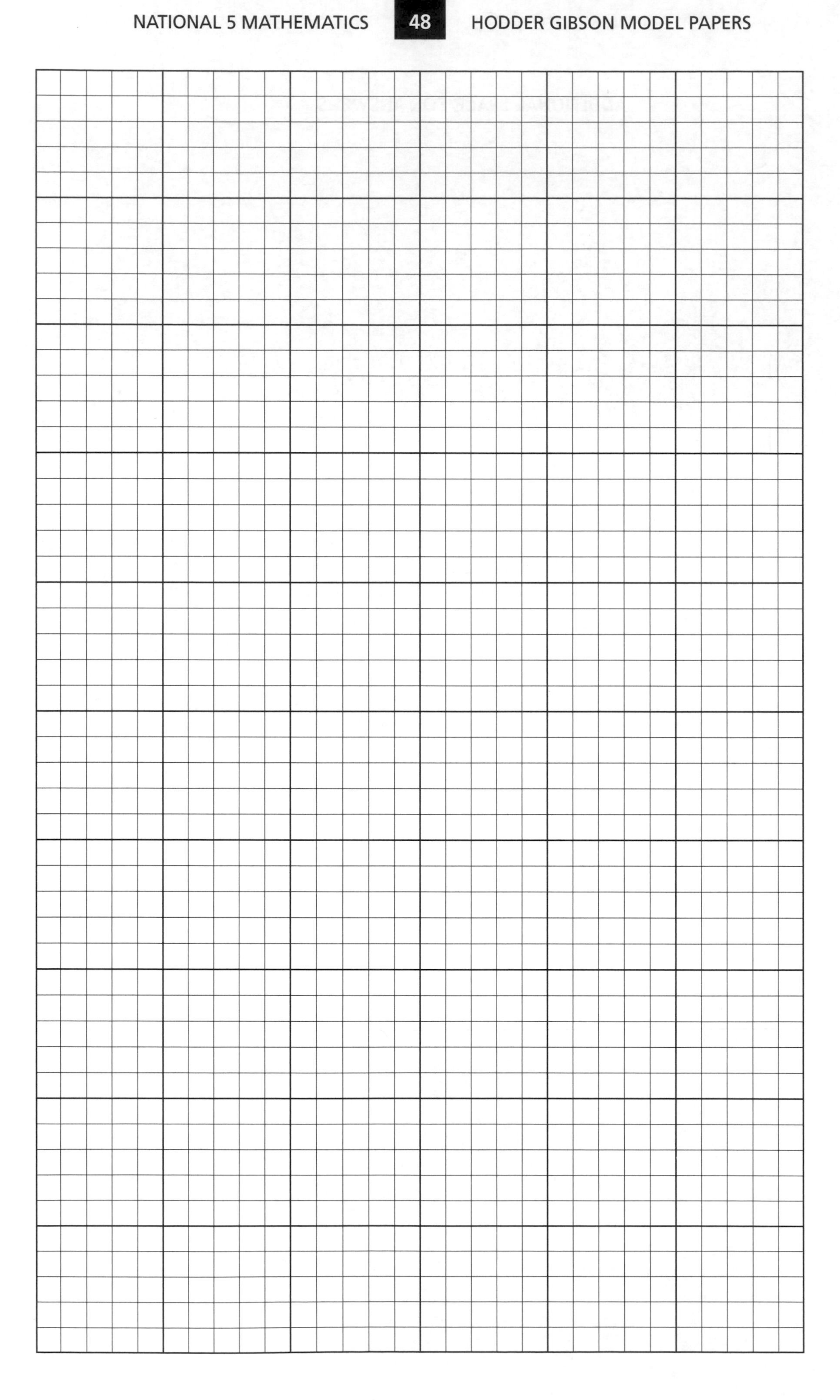

Mathematics
Paper 2

Duration — 1 hour and 30 minutes

Total marks — 50

You may use a calculator.

Attempt ALL questions.

Use **blue** or **black** ink. Pencil may be used for graphs and diagrams only.

Write your working and answers in the spaces provided. Additional space for answers is provided at the end of this booklet. If you use this space, write clearly the number of the question you are attempting.

Square-ruled paper is provided at the back of this booklet.

Full credit will be given only to solutions which contain appropriate working.

State the units for your answer where appropriate.

Before leaving the examination room you must give this booklet to the Invigilator. If you do not, you may lose all the marks for this paper.

FORMULAE LIST

The roots of $ax^2 + bx + c = 0$ are $x = \dfrac{-b \pm \sqrt{(b^2 - 4ac)}}{2a}$

Sine rule: $\dfrac{a}{\sin A} = \dfrac{b}{\sin B} = \dfrac{c}{\sin C}$

Cosine rule: $a^2 = b^2 + c^2 - 2bc\cos A$ or $\cos A = \dfrac{b^2 + c^2 - a^2}{2bc}$

Area of a triangle: $A = \frac{1}{2}ab\sin C$

Volume of a sphere: $V = \frac{4}{3}\pi r^3$

Volume of a cone: $V = \frac{1}{3}\pi r^2 h$

Volume of a pyramid: $V = \frac{1}{3}Ah$

Standard deviation: $s = \sqrt{\dfrac{\Sigma(x - \bar{x})^2}{n-1}} = \sqrt{\dfrac{\Sigma x^2 - (\Sigma x)^2/n}{n-1}}$, where n is the sample size.

MARKS | DO NOT WRITE IN THIS MARGIN

1. Alistair buys an antique chair for £600.

   It is expected to increase in value at the rate of 4·5% each year.

   How much is it expected to be worth in 3 years? **3**

2. A rugby team scored the following points in a series of matches.

   13 7 0 9 7 8 5

   (a) For this sample calculate the mean and the standard deviation. **3**

   The following season the team appoints a new coach.

   A similar series of matches produces a mean of 27 and a standard deviation of 3·25.

   (b) Make two valid comparisons about the performance of the team under the new coach. **2**

   Total marks 5

MARKS | DO NOT WRITE IN THIS MARGIN

3. The diagram shows a cuboid OPQR,STUV relative to the coordinate axes.

P is the point (4, 0, 0), Q is (4, 2, 0) and U is (4, 2, 3).

M is the midpoint of OR.

N is the point on UQ such that UN = $\frac{1}{3}$ UQ.

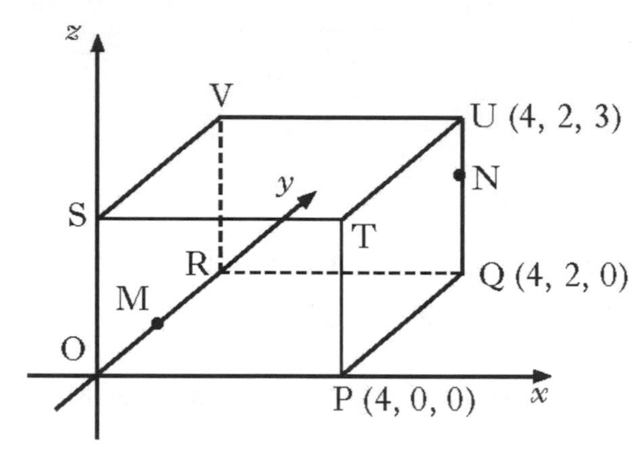

State the coordinates of M and N. **2**

4. Find the equation of the straight line shown in the diagram.

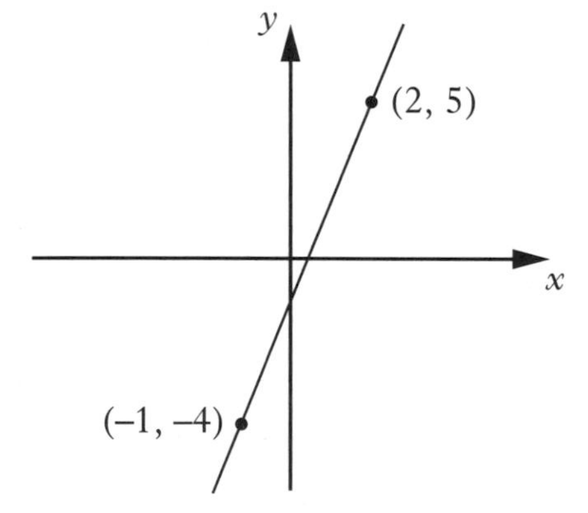

3

MARKS DO NOT WRITE IN THIS MARGIN

5. A spiral staircase is being designed.

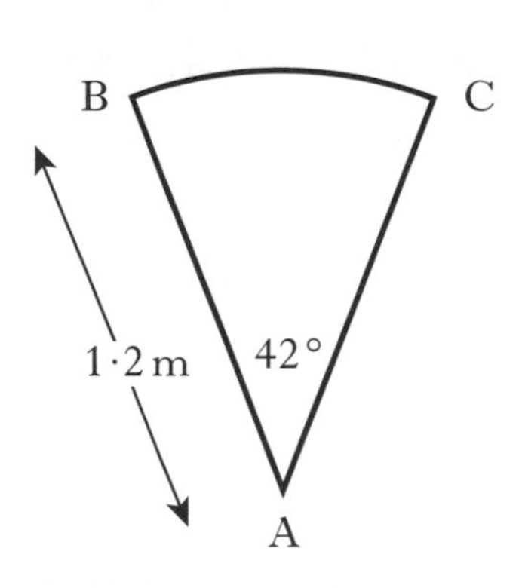

Each step is made from a sector of a circle as shown.

The radius is 1·2 metres.

Angle BAC is 42°.

For the staircase to pass safety regulations, the arc BC must be at least 0·9 metres.

Will the staircase pass safety regulations? **4**

6. A glass ornament is in the shape of a cone partly filled with coloured water.

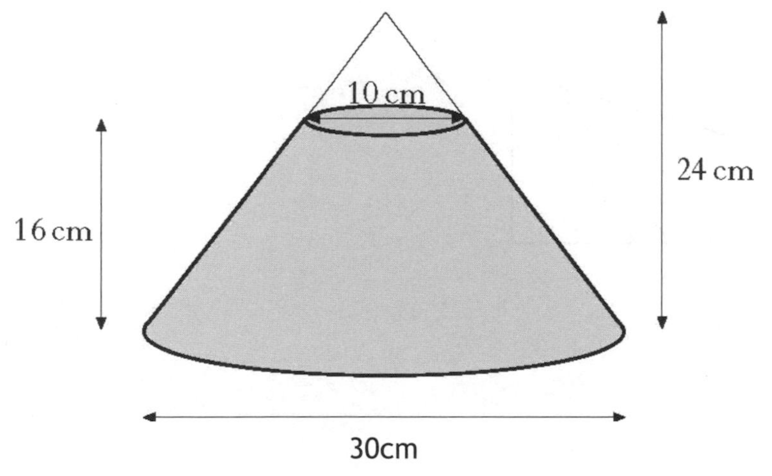

The cone is 24 centimetres high and has a base of diameter 30 centimetres.

The water is 16 centimetres deep and measures 10 centimetres across the top.

What is the volume of the water?

Give your answer **correct to 2 significant figures**. **5**

MARKS DO NOT WRITE IN THIS MARGIN

7. The price for Paul's summer holiday is £894·40.

The price includes a 4% booking fee.

What is the price of his holiday without the booking fee? **3**

8. A heavy metal beam, AB, rests against a vertical wall as shown.

The length of the beam is 8 metres and it makes and angle of 59° with the ground.

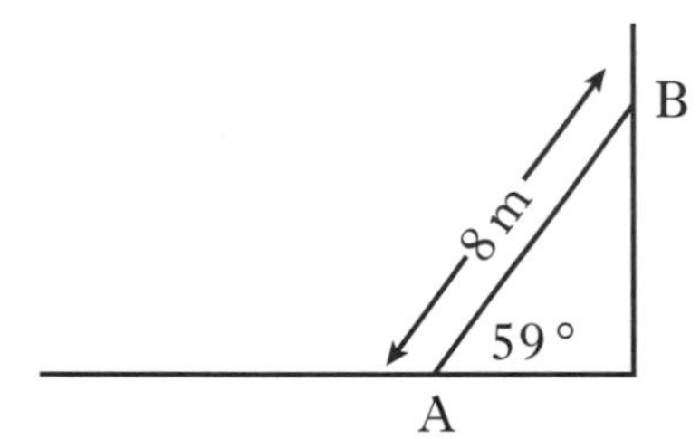

A cable, CB, is fixed to the ground at C and is attached to the top of the beam at B.

The cable makes an angle of 22° with the ground.

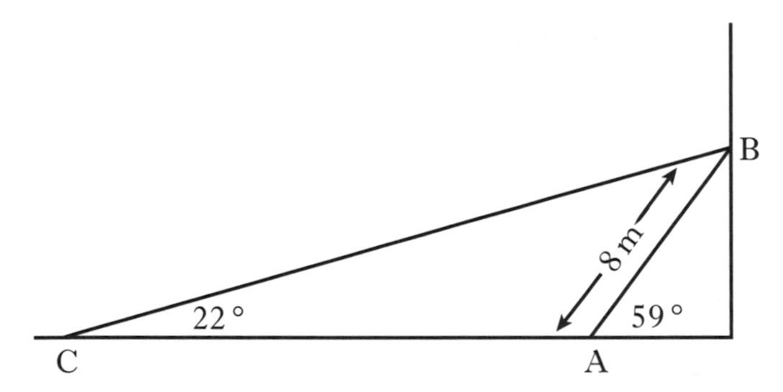

Calculate the length of cable CB. **4**

MARKS DO NOT WRITE IN THIS MARGIN

9. A necklace is made of beads which are mathematically similar.

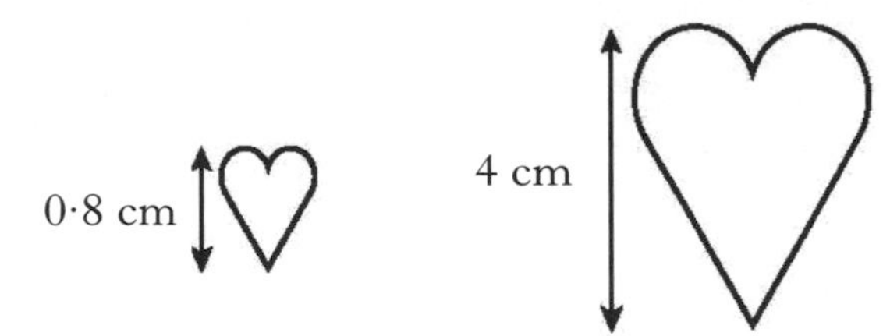

The height of the smaller bead is 0·8 centimetres and its area is 0·6 square centimetres.
The height of the larger bead is 4 centimetres.
Find the area of the larger bead. **3**

10. Paving stones are in the shape of a rhombus.

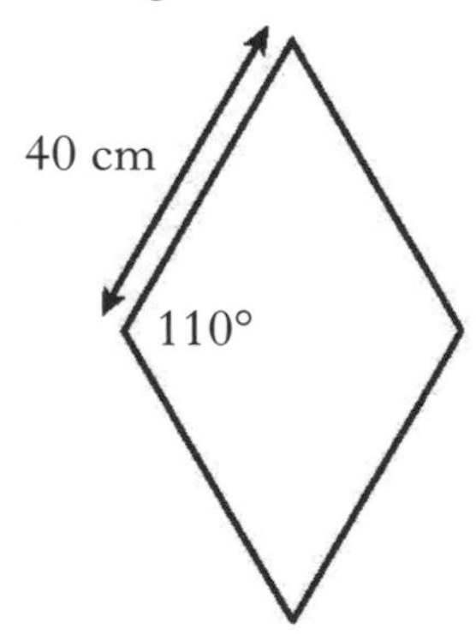

The side of each rhombus is 40 centimetres long.

The obtuse angle is 110°.

Find the area of one paving stone. **4**

MARKS DO NOT WRITE IN THIS MARGIN

11. $f(x) = 3\sin x^o,\ \ 0 \leq x \leq 360$.

(a) Find $f(270)$. **1**

(b) $f(t) = 0 \cdot 6$.

Find the two possible values of t. **4**

Total marks 5

12. A tanker delivers oil to garages.

The tanker has a circular cross-section as shown in the diagram below.

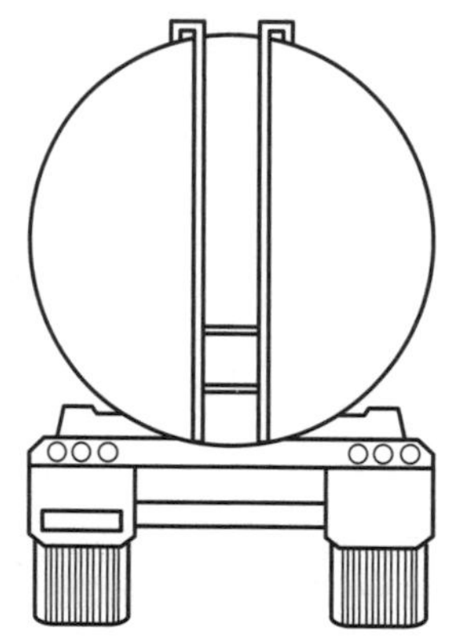

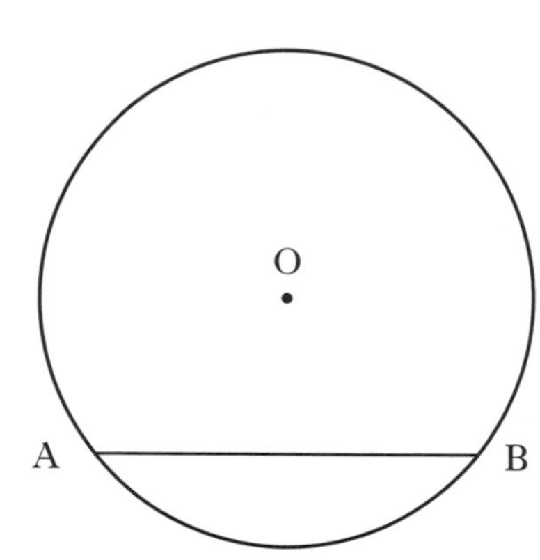

The radius of the circle, O, is 1·9 metres.

The width of the surface of the oil, represented by AB in the diagram, is 2·2 metres.

Calculate the depth of the oil in the tanker. **4**

MARKS | DO NOT WRITE IN THIS MARGIN

13. Triangles PQR and STU are mathematically similar.

The scale factor is 3 and PR corresponds to SU.

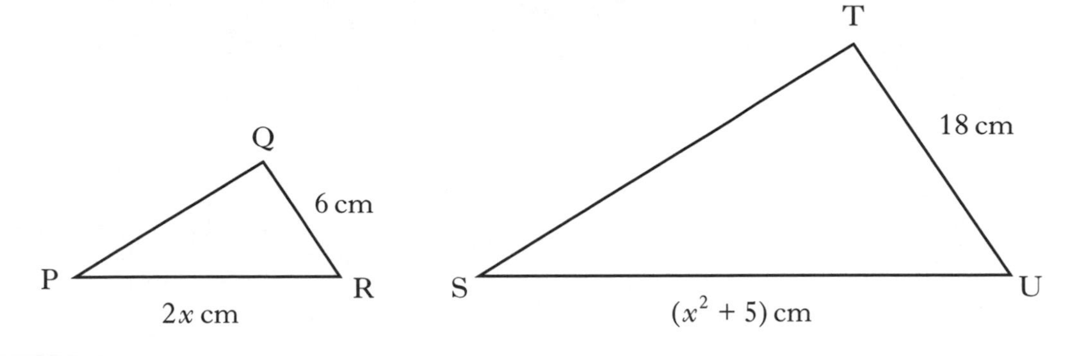

(a) Show that $x^2 - 6x + 5 = 0$. **2**

(b) Given that QR is the shortest side of triangle PQR, find the value of x. **3**

Total marks 5

[END OF MODEL PRACTICE PAPER]

ADDITIONAL SPACE FOR ANSWERS

ADDITIONAL SPACE FOR ANSWERS

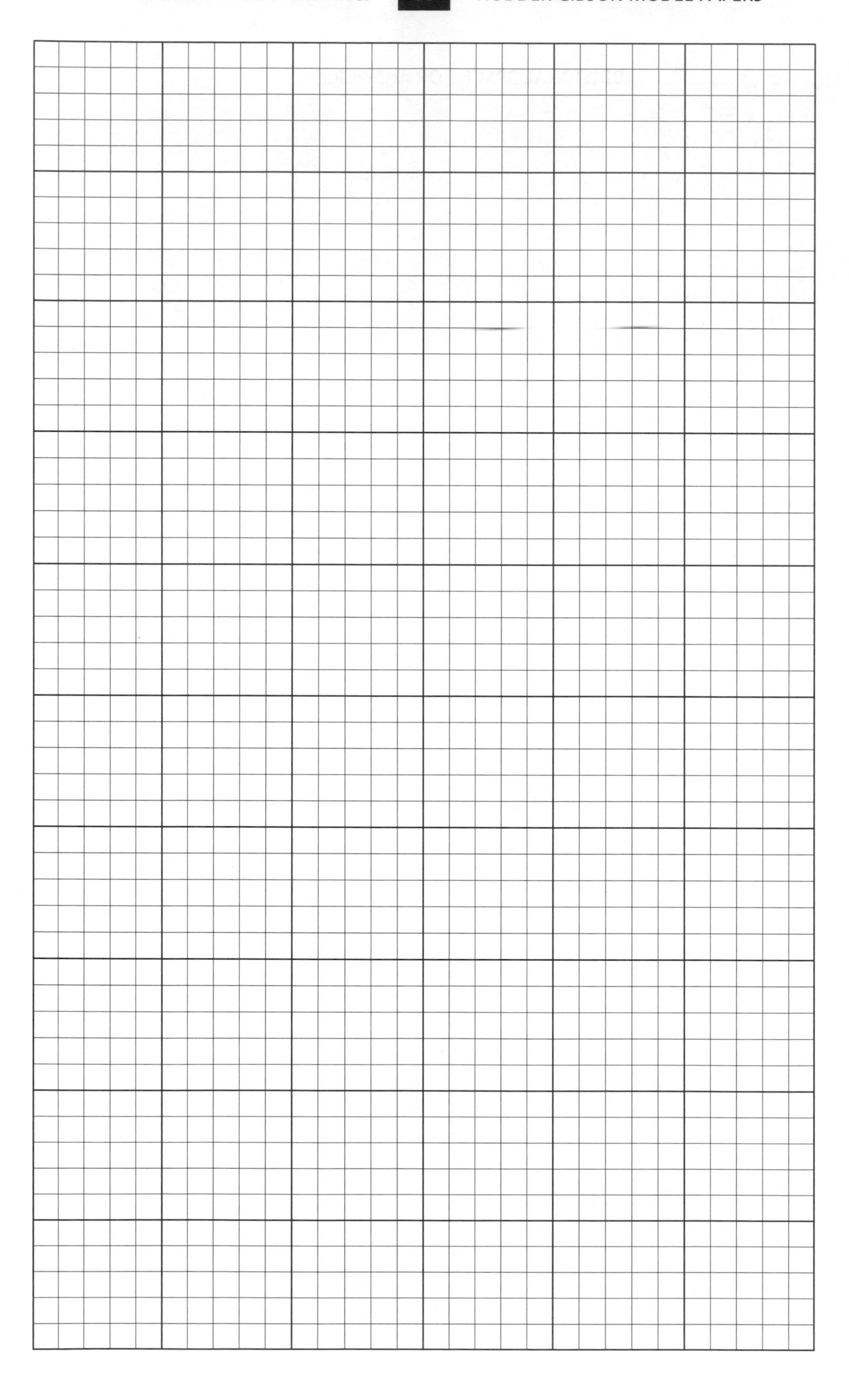

NATIONAL 5

Model Paper 2

Whilst this Model Practice Paper has been specially commissioned by Hodder Gibson for use as practice for the National 5 exams, the key reference documents remain the SQA Specimen Paper 2013 and the SQA Past Paper 2014.

Mathematics
Paper 1
(Non-Calculator)

Duration — 1 hour

Total marks — 40

You may NOT use a calculator.

Attempt ALL questions.

Use **blue** or **black** ink. Pencil may be used for graphs and diagrams only.

Write your working and answers in the spaces provided. Additional space for answers is provided at the end of this booklet. If you use this space, write clearly the number of the question you are attempting.

Square-ruled paper is provided at the back of this booklet.

Full credit will be given only to solutions which contain appropriate working.

State the units for your answer where appropriate.

Before leaving the examination room you must give this booklet to the Invigilator.
If you do not, you may lose all the marks for this paper.

FORMULAE LIST

The roots of $ax^2 + bx + c = 0$ are $x = \dfrac{-b \pm \sqrt{(b^2 - 4ac)}}{2a}$

Sine rule: $\dfrac{a}{\sin A} = \dfrac{b}{\sin B} = \dfrac{c}{\sin C}$

Cosine rule: $a^2 = b^2 + c^2 - 2bc\cos A$ or $\cos A = \dfrac{b^2 + c^2 - a^2}{2bc}$

Area of a triangle: $A = \frac{1}{2}ab\sin C$

Volume of a sphere: $V = \frac{4}{3}\pi r^3$

Volume of a cone: $V = \frac{1}{3}\pi r^2 h$

Volume of a pyramid: $V = \frac{1}{3}Ah$

Standard deviation: $s = \sqrt{\dfrac{\Sigma(x - \overline{x})^2}{n - 1}} = \sqrt{\dfrac{\Sigma x^2 - (\Sigma x)^2/n}{n - 1}}$, where n is the sample size.

MARKS | DO NOT WRITE IN THIS MARGIN

1. Evaluate $\frac{2}{5} \div 1\frac{1}{10}$. **2**

2. Factorise fully $2m^2 - 18$. **2**

3. Given that $f(x) = 5 - x^2$, evaluate $f(-3)$. **2**

MARKS | DO NOT WRITE IN THIS MARGIN

4. Solve the equation $3x+1=\dfrac{x-5}{2}$. **3**

5. Express $\sqrt{63}+\sqrt{28}-\sqrt{7}$ as a surd in its simplest form. **3**

MARKS | DO NOT WRITE IN THIS MARGIN

6. Express $x^2 + 10x + 17$ in the form $(x + p)^2 + q$. **2**

7. Alan is taking part in a quiz. He is awarded x points for each correct answer and y points for each wrong answer. During the quiz, Alan gets 24 questions correct and 6 wrong. He scores 60 points.

(a) Write down an equation in x and y which satisfies the above condition. **1**

Helen also takes part in the quiz. She gets 20 questions correct and 10 wrong. She scores 40 points.

(b) Write down a second equation in x and y which satisfies this condition. **1**

(c) Calculate the score for David who gets 17 correct and 13 wrong. **4**

Total marks 6

MARKS DO NOT WRITE IN THIS MARGIN

8. A circle, centre O, is shown below.

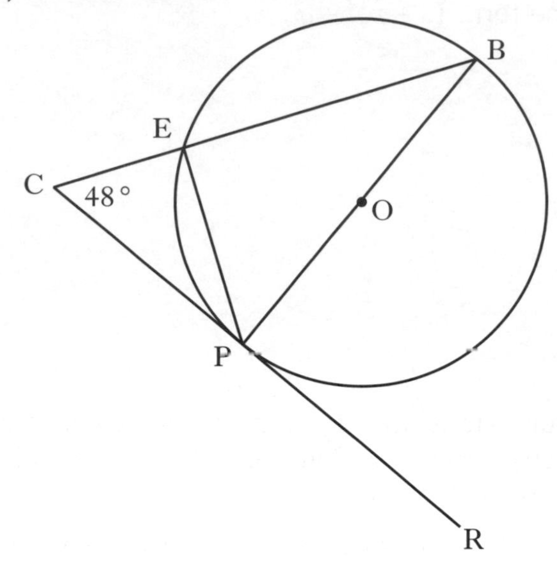

In the circle

- PB is a diameter
- CR is a tangent to the circle at point P
- Angle BCP is 48°.

Calculate the size of EPR. **3**

MARKS DO NOT WRITE IN THIS MARGIN

9. The graph below shows the relationship between the History and Geography marks of a class of students.

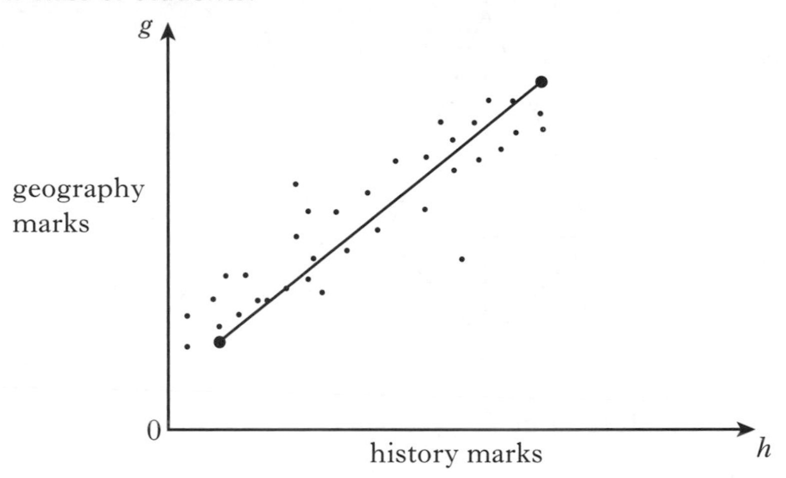

A best-fitting straight line, AB has been drawn.

Point A represents 12 marks for history and 20 marks for geography.

Point B represents 92 marks for history and 80 marks for geography.

Find the equation of the straight line AB in terms of h and g. **4**

MARKS | DO NOT WRITE IN THIS MARGIN

10. A kite PQRS is shown below.

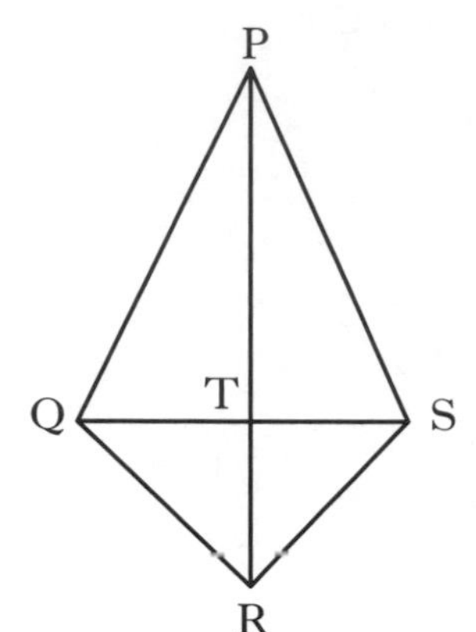

The diagonals of the kite intersect at T.

PT = 2TR.

$\overrightarrow{PR}$ represents vector **a**.

$\overrightarrow{QS}$ represents vector **b**.

Express $\overrightarrow{PS}$ in terms of **a** and **b**. **2**

MARKS | DO NOT WRITE IN THIS MARGIN

11. In the triangle ABC

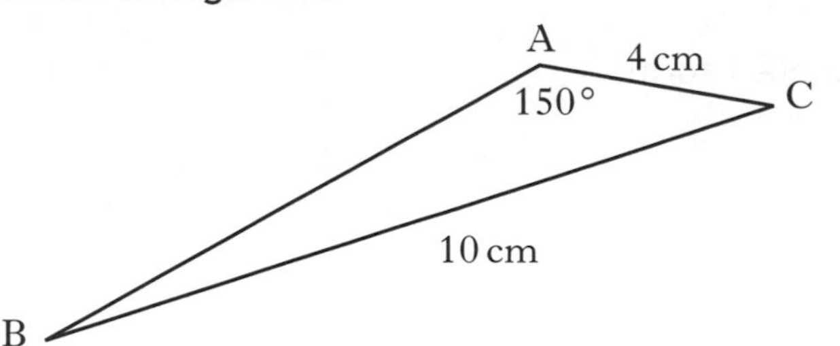

- AC = 4 centimetres
- BC = 10 centimetres
- Angle BAC = 150°.

Given that $\sin 30^o = \frac{1}{2}$, show that $\sin B = \frac{1}{5}$. **4**

12. Express $\frac{b^{\frac{1}{2}} \times b^{\frac{5}{2}}}{b^2}$ in its simplest form. **2**

MARKS DO NOT WRITE IN THIS MARGIN

13. Express $\frac{5p^2}{8} \div \frac{p}{2}$ as a fraction in its simplest form. **3**

14. Prove that $\frac{\sin^2 A}{1-\sin^2 A} = \tan^2 A$. **2**

[END OF MODEL PRACTICE PAPER]

ADDITIONAL SPACE FOR ANSWERS

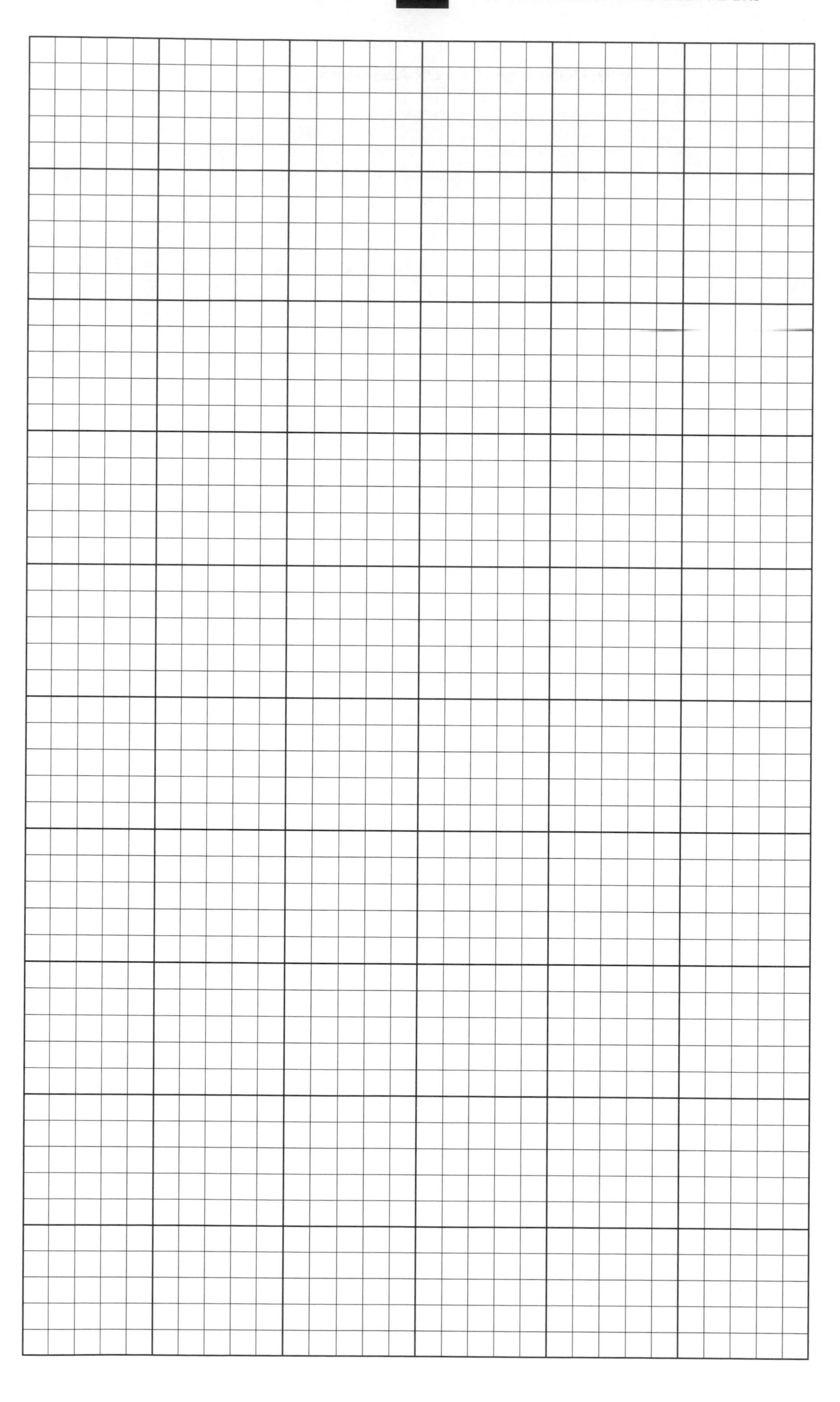

National
Qualifications
MODEL PAPER 2

Mathematics
Paper 2

Duration — 1 hour and 30 minutes

Total marks — 50

You may use a calculator.

Attempt ALL questions.

Use **blue** or **black** ink. Pencil may be used for graphs and diagrams only.

Write your working and answers in the spaces provided. Additional space for answers is provided at the end of this booklet. If you use this space, write clearly the number of the question you are attempting.

Square-ruled paper is provided at the back of this booklet.

Full credit will be given only to solutions which contain appropriate working.

State the units for your answer where appropriate.

Before leaving the examination room you must give this booklet to the Invigilator.
If you do not, you may lose all the marks for this paper.

FORMULAE LIST

The roots of $ax^2 + bx + c = 0$ are $x = \dfrac{-b \pm \sqrt{(b^2 - 4ac)}}{2a}$

Sine rule: $\dfrac{a}{\sin A} = \dfrac{b}{\sin B} = \dfrac{c}{\sin C}$

Cosine rule: $a^2 = b^2 + c^2 - 2bc\cos A$ or $\cos A = \dfrac{b^2 + c^2 - a^2}{2bc}$

Area of a triangle: $A = \frac{1}{2}ab\sin C$

Volume of a sphere: $V = \frac{4}{3}\pi r^3$

Volume of a cone: $V = \frac{1}{3}\pi r^2 h$

Volume of a pyramid: $V = \frac{1}{3}Ah$

Standard deviation: $s = \sqrt{\dfrac{\Sigma(x - \overline{x})^2}{n-1}} = \sqrt{\dfrac{\Sigma x^2 - (\Sigma x)^2/n}{n-1}}$, where n is the sample size.

MARKS DO NOT WRITE IN THIS MARGIN

1. $E = mc^2$.

Find the value of E when $m = 3{\cdot}6 \times 10^{-2}$ and $c = 3 \times 10^8$.

Give your answer **in scientific notation.** 3

2. Expand fully and simplify $x(x-1)^2$. 2

MARKS | DO NOT WRITE IN THIS MARGIN

3. A sector of a circle, centre O, is shown below.

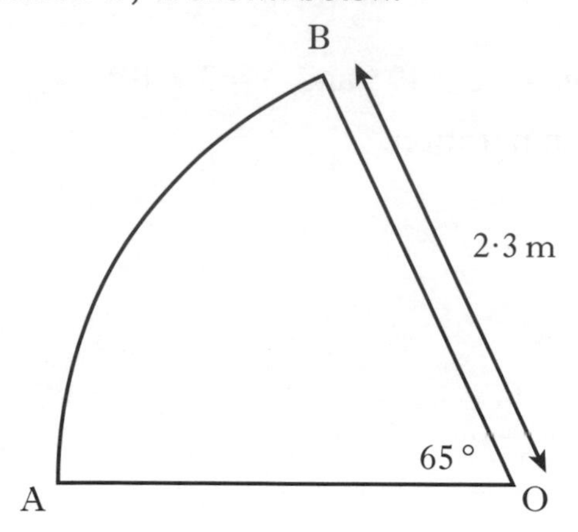

The radius of the circle is 2·3 metres.

Angle AOB is 65°.

Find the length of the arc AB. **3**

4. Change the subject of the formula $p = q + 2r^2$ to r. **3**

MARKS | DO NOT WRITE IN THIS MARGIN

5. Solve the equation $2x^2 + 3x - 7 = 0$.

Give your answer **correct to 2 significant figures.** 4

6. The marks of a group of students in their October test are listed below.

41 56 68 59 43 37 70 58 61 47 75 66

(a) Calculate the median and the interquartile range. 3

The teacher arranges extra homework classes for the students before the next test in December.

In this test, the median is 67 and the interquartile range is 14.

(b) Make **two** appropriate comments comparing the marks in the October and December tests. 2

Total marks 5

MARKS | DO NOT WRITE IN THIS MARGIN

7. Two yachts leave from harbour H.

Yacht A sails on a bearing of 072° for 30 kilometres and stops.

Yacht B sails on a bearing of 140° for 50 kilometres and stops.

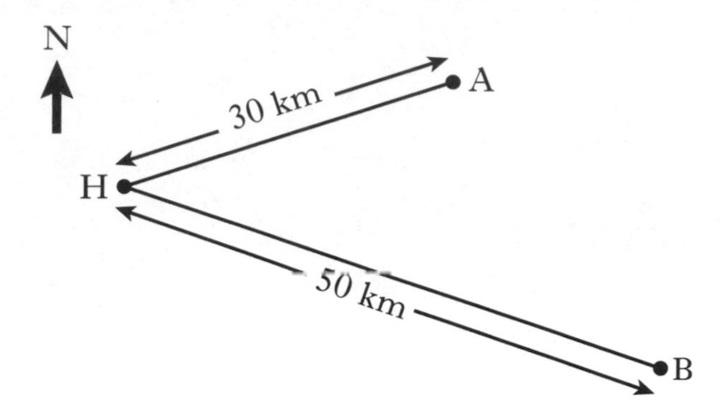

How far apart are the two yachts when they have both stopped?

Do not use a scale drawing. 4

8. Two rectangular solar panels, A and B, are mathematically similar.

Panel A has a diagonal of 90 centimetres and an area of 4020 square centimetres.

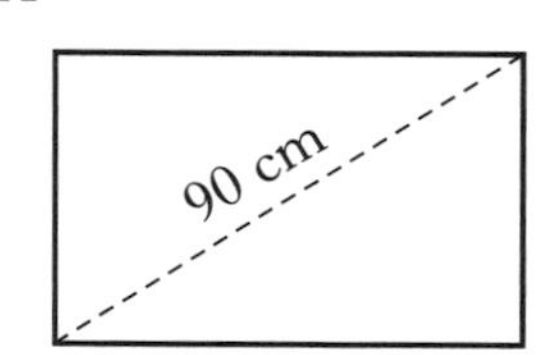

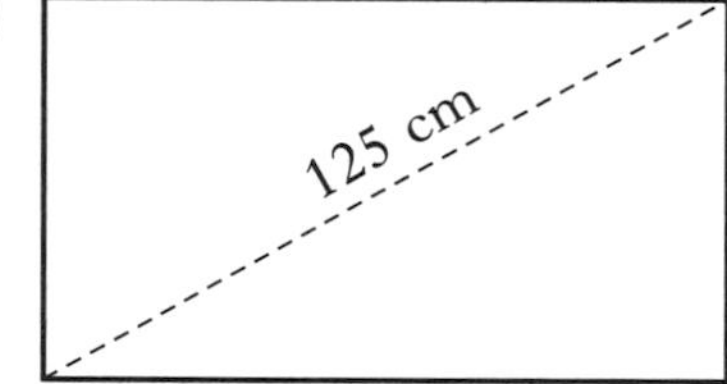

A salesman claims that panel B, with a diagonal of 125 centimetres, will be double the area of panel A.

Is this claim justified?

Show all your working. 4

MARKS | DO NOT WRITE IN THIS MARGIN

9. Vector **u** has components $\begin{pmatrix} 2 \\ 0 \\ 1 \end{pmatrix}$ and vector **v** has components $\begin{pmatrix} 1 \\ 2 \\ -4 \end{pmatrix}$.

Calculate the magnitude of $2\mathbf{u} - \mathbf{v}$. **2**

10. A triangular paving slab has measurements as shown.

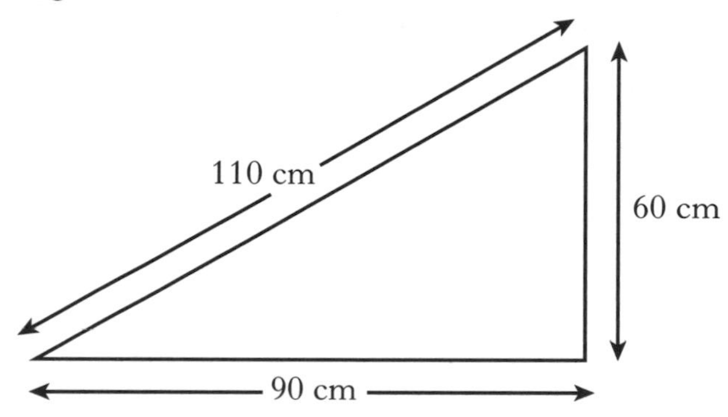

Is the slab in the shape of a right-angled triangle?

Show all your working. **3**

MARKS | DO NOT WRITE IN THIS MARGIN

11. The diagram below shows a pyramid.

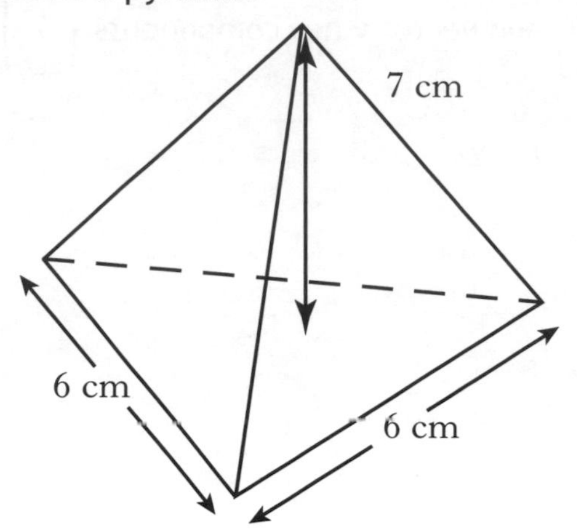

The base of the pyramid is an equilateral triangle of side 6 centimetres.

The height of the pyramid is 7 centimetres.

Calculate the volume of the pyramid. **3**

MARKS | DO NOT WRITE IN THIS MARGIN

12. The graph below shows part of a parabola with equation of the form

$$y = (x + a)^2 + b\,.$$

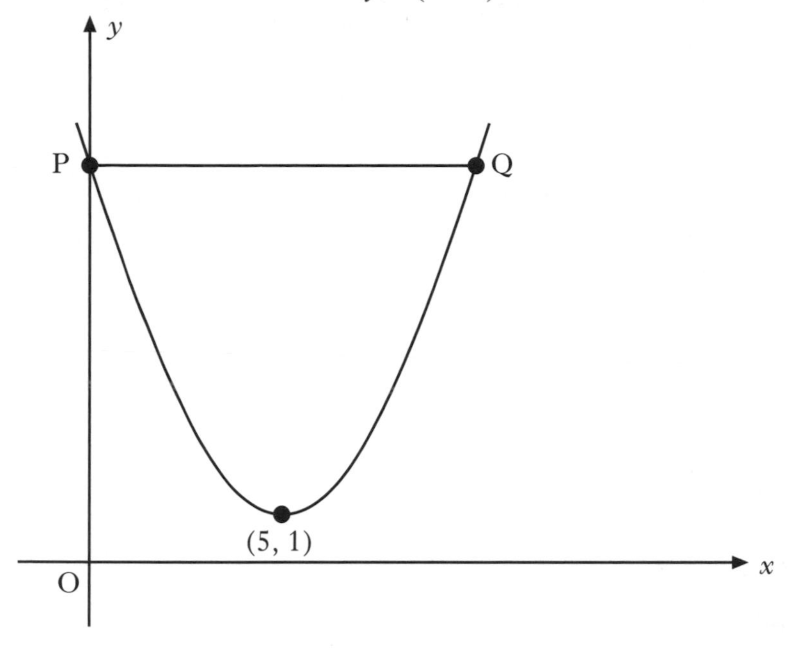

(a) State the values of a and b. **2**

(b) The line PQ is parallel to the x-axis.

Find the coordinates of points P and Q. **3**

Total marks 5

MARKS | DO NOT WRITE IN THIS MARGIN

13. Part of the graph of $y = 4\sin x° - 3$ is shown below.

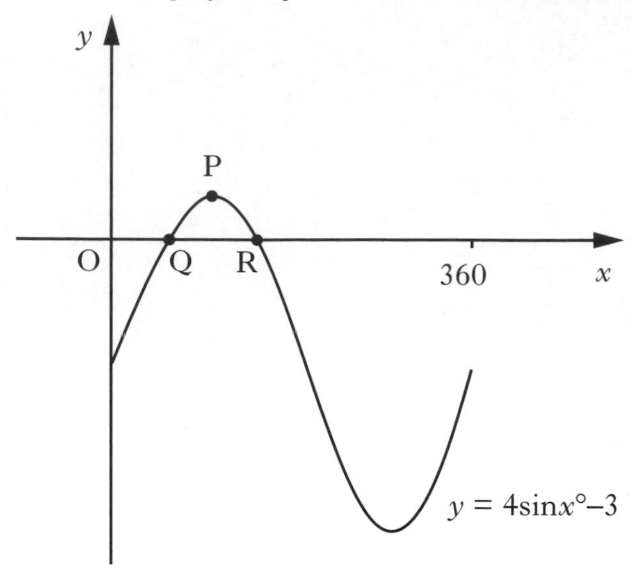

The graph cuts the x-axis at Q and R.

P is the maximum turning point.

(a) Write down the coordinates of P. **1**

(b) Calculate the x-coordinates of Q and R. **4**

Total marks 5

MARKS DO NOT WRITE IN THIS MARGIN

14. The diagram shows the path of a flare after it is fired.

The height, h metres above sea level, of the flare is given by

$h = 48 + 8t - t^2$ where t is the number of seconds after firing.

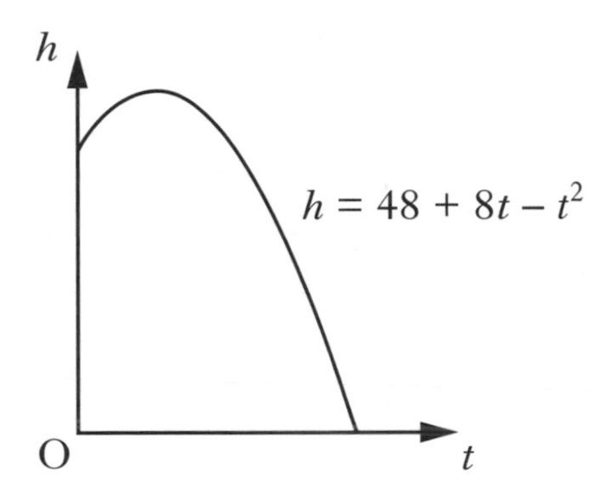

Calculate, **algebraically**, the time taken for the flare to enter the sea. **4**

[END OF MODEL PRACTICE PAPER]

ADDITIONAL SPACE FOR ANSWERS

ADDITIONAL SPACE FOR ANSWERS

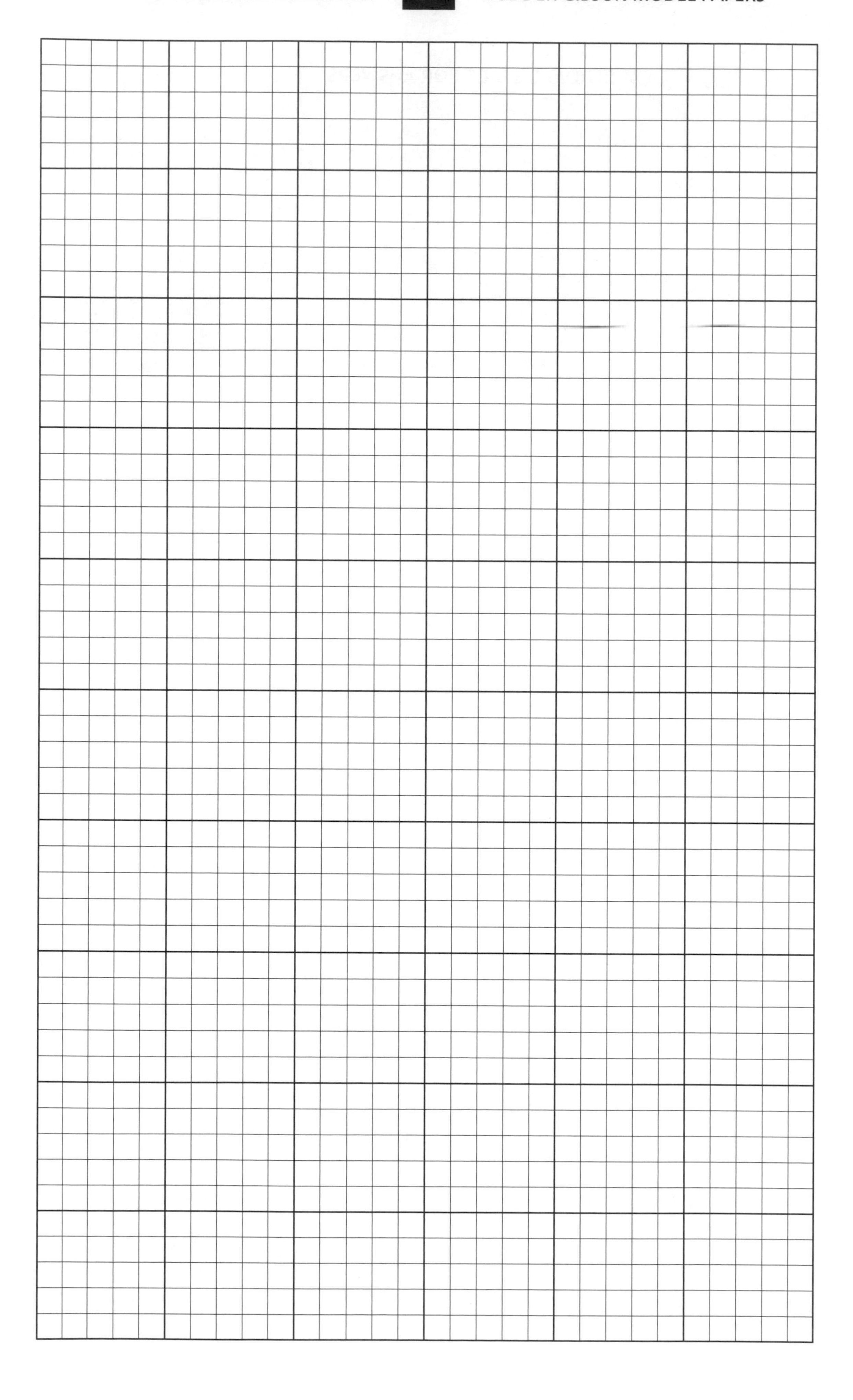

NATIONAL 5

Model Paper 3

Whilst this Model Practice Paper has been specially commissioned by Hodder Gibson for use as practice for the National 5 exams, the key reference documents remain the SQA Specimen Paper 2013 and the SQA Past Paper 2014.

Mathematics
Paper 1
(Non-Calculator)

Duration — 1 hour

Total marks — 40

You may NOT use a calculator.

Attempt ALL questions.

Use **blue** or **black** ink. Pencil may be used for graphs and diagrams only.

Write your working and answers in the spaces provided. Additional space for answers is provided at the end of this booklet. If you use this space, write clearly the number of the question you are attempting.

Square-ruled paper is provided at the back of this booklet.

Full credit will be given only to solutions which contain appropriate working.

State the units for your answer where appropriate.

Before leaving the examination room you must give this booklet to the Invigilator.
If you do not, you may lose all the marks for this paper.

FORMULAE LIST

The roots of $ax^2 + bx + c = 0$ are $x = \dfrac{-b \pm \sqrt{(b^2 - 4ac)}}{2a}$

Sine rule: $\dfrac{a}{\sin A} = \dfrac{b}{\sin B} = \dfrac{c}{\sin C}$

Cosine rule: $a^2 = b^2 + c^2 - 2bc\cos A$ or $\cos A = \dfrac{b^2 + c^2 - a^2}{2bc}$

Area of a triangle: $A = \frac{1}{2}ab\sin C$

Volume of a sphere: $V = \frac{4}{3}\pi r^3$

Volume of a cone: $V = \frac{1}{3}\pi r^2 h$

Volume of a pyramid: $V = \frac{1}{3}Ah$

Standard deviation: $s = \sqrt{\dfrac{\Sigma(x-\bar{x})^2}{n-1}} = \sqrt{\dfrac{\Sigma x^2 - (\Sigma x)^2/n}{n-1}}$, where n is the sample size.

MARKS | DO NOT WRITE IN THIS MARGIN

1. Evaluate $2\frac{1}{3}+\frac{5}{6} of 1\frac{2}{5}$. **3**

2. Solve the inequality $5-x>2(x+1)$. **2**

3. Factorise $2p^2-5p-12$. **2**

MARKS DO NOT WRITE IN THIS MARGIN

4. The temperature, in degrees Celsius, at mid-day in a seaside town and the sales, in pounds, of umbrellas are shown in the scattergraph below.

A line of best fit has been drawn.

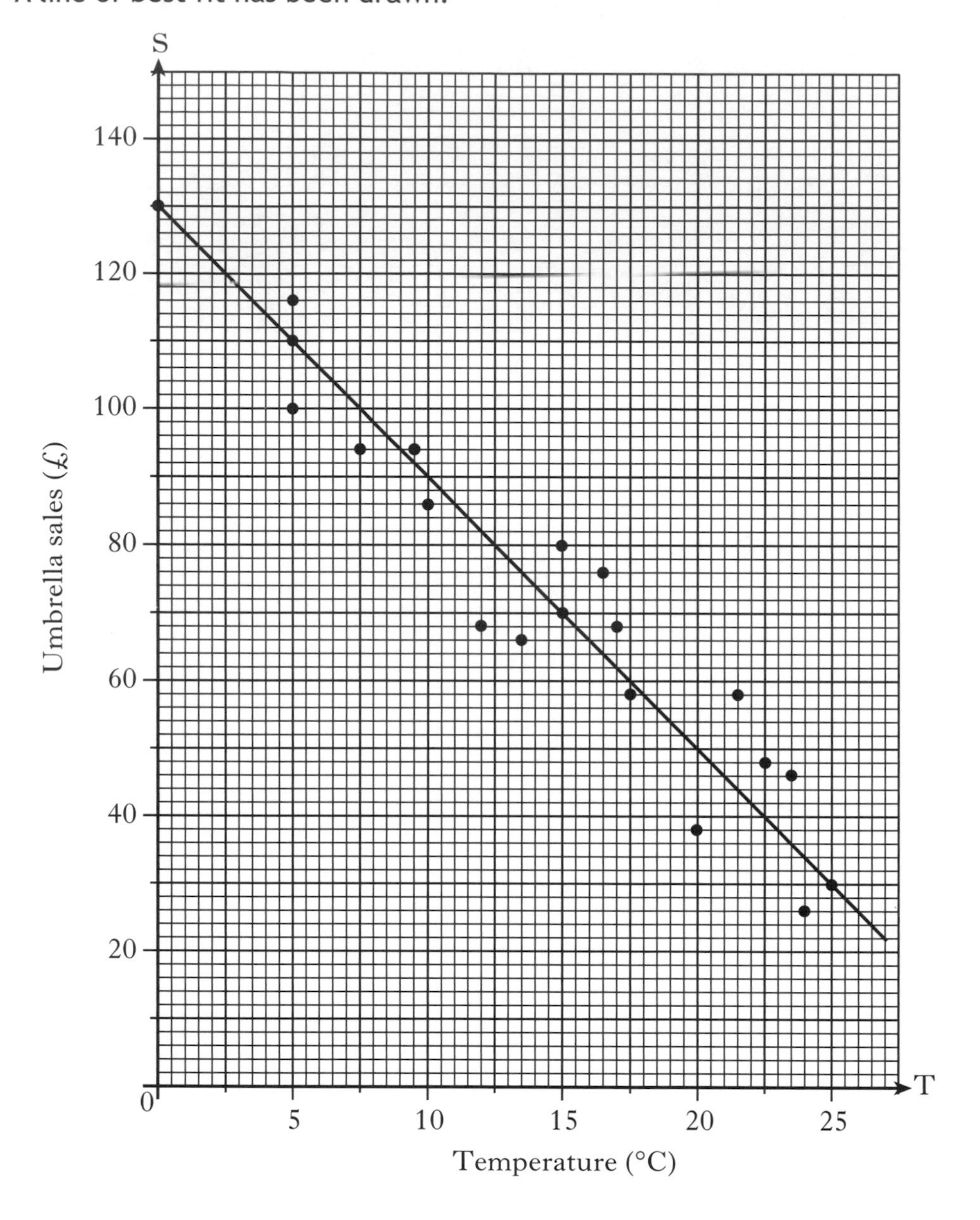

(a) Find the equation of the line of best fit. **3**

(b) **Use your answer to part (a)** to predict the sales for a day when the temperature is 30 degrees Celsius. **1**

Total marks 4

MARKS DO NOT WRITE IN THIS MARGIN

5. The diagram below represents a sphere.

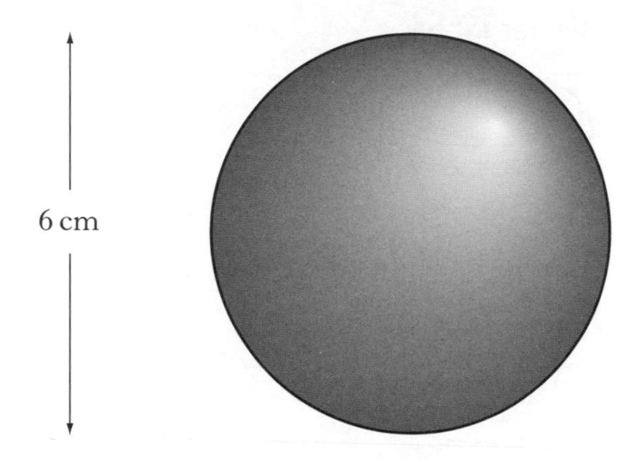

The sphere has a diameter of 6 centimetres.

Calculate its volume.

Take π = 3·14. **2**

6. Solve algebraically the system of equations

$$2x - 5y = 24$$

$$7x + 8y = 33.$$

3

MARKS DO NOT WRITE IN THIS MARGIN

7. Coffee is sold in regular cups and large cups.

The two cups are mathematically similar in shape.

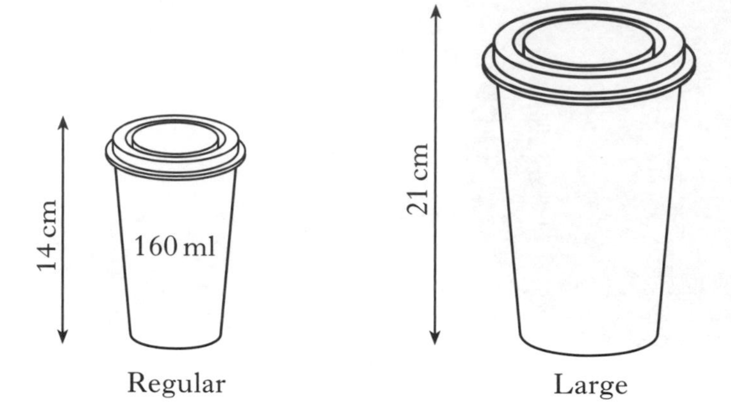

The regular cup is 14 centimetres high and hold 160 millilitres.

The large cup is 21 centimetres high.

Calculate how many millilitres the large cup holds. **4**

8. (a) Show that the standard deviation of 1, 1, 1, 2 and 5 is $\sqrt{3}$. **3**

(b) **Write down** the standard deviation of 101, 101, 101, 102 and 105. **1**

Total marks 4

MARKS | DO NOT WRITE IN THIS MARGIN

9. Cleano washing powder is on a special offer.

Each box on special offer contains 20% more powder than the standard box.

A box on special offer contains 900 grams of powder.

How many grams of powder does the standard box hold? **3**

10. The graph shown below has an equation of the form $y = \cos(x - a)^{o}$.

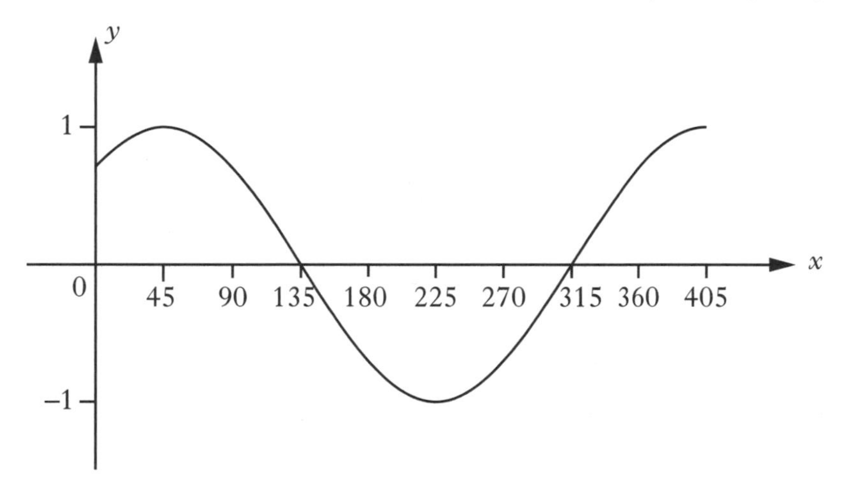

Write down the value of a. **1**

MARKS | DO NOT WRITE IN THIS MARGIN

11. Express $\frac{12}{\sqrt{2}}$ with a rational denominator.

Give your answer in its simplest form. **2**

12. Each day, Marissa drives 40 kilometres to work.

(a) On Monday, she drives at a speed of x kilometres per hour.

Find the time taken, in terms of x, for her journey. **1**

(b) On Tuesday, she drives 5 kilometres per hour faster.

Find the time taken, in terms of x, for this journey. **1**

(c) Hence find an expression, in terms of x, for the difference in times of the two journeys. **3**

Total marks 5

MARKS DO NOT WRITE IN THIS MARGIN

13. William Watson's fast Foods use a logo based on parts of three identical parabola.

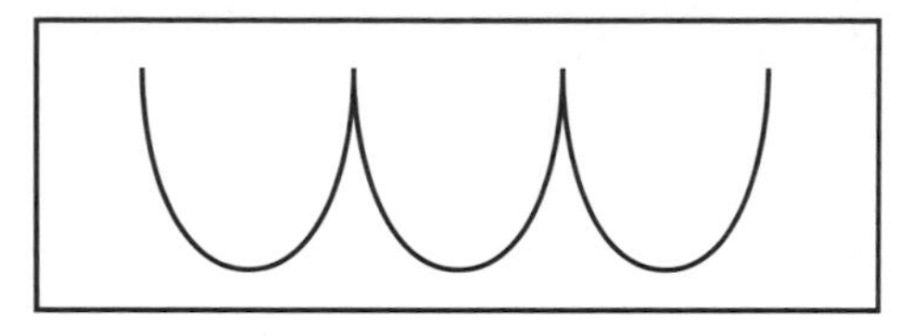

This logo is represented in the diagram below.

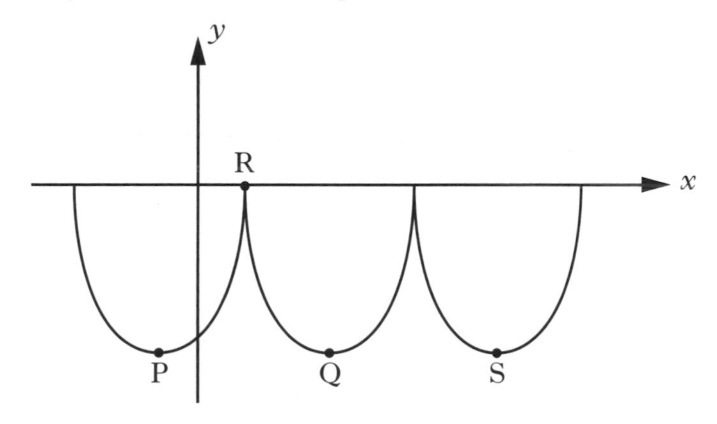

The first parabola has turning point P and equation $y = (x + 2)^2 - 16$.

(a) State the coordinates of P. **2**

(b) If R is the point (2,0), find the coordinates of Q, the minimum turning point of the second parabola. **1**

(c) Find the equation of the parabola with turning point S. **2**

Total marks 5

[END OF MODEL PRACTICE PAPER]

ADDITIONAL SPACE FOR ANSWERS

ADDITIONAL SPACE FOR ANSWERS

Mathematics
Paper 2

Duration — 1 hour and 30 minutes

Total marks — 50

You may use a calculator.

Attempt ALL questions.

Use **blue** or **black** ink. Pencil may be used for graphs and diagrams only.

Write your working and answers in the spaces provided. Additional space for answers is provided at the end of this booklet. If you use this space, write clearly the number of the question you are attempting.

Square-ruled paper is provided at the back of this booklet.

Full credit will be given only to solutions which contain appropriate working.

State the units for your answer where appropriate.

Before leaving the examination room you must give this booklet to the Invigilator.
If you do not, you may lose all the marks for this paper.

FORMULAE LIST

The roots of $ax^2 + bx + c = 0$ are $x = \frac{-b \pm \sqrt{(b^2 - 4ac)}}{2a}$

Sine rule: $\frac{a}{\sin A} = \frac{b}{\sin B} = \frac{c}{\sin C}$

Cosine rule: $a^2 = b^2 + c^2 - 2bc\cos A$ or $\cos A = \frac{b^2 + c^2 - a^2}{2bc}$

Area of a triangle: $A = \frac{1}{2}ab\sin C$

Volume of a sphere: $V = \frac{4}{3}\pi r^3$

Volume of a cone: $V = \frac{1}{3}\pi r^2 h$

Volume of a pyramid: $V = \frac{1}{3}Ah$

Standard deviation: $s = \sqrt{\frac{\Sigma(x - \bar{x})^2}{n-1}} = \sqrt{\frac{\Sigma x^2 - (\Sigma x)^2/n}{n-1}}$, where n is the sample size.

MARKS | DO NOT WRITE IN THIS MARGIN

1. The National Debt of the United Kingdom was recently calculated as

£1 157 818 887 139.

Round this amount to **four significant figures**. 1

2. The diagram shows vectors **s** and **t**.

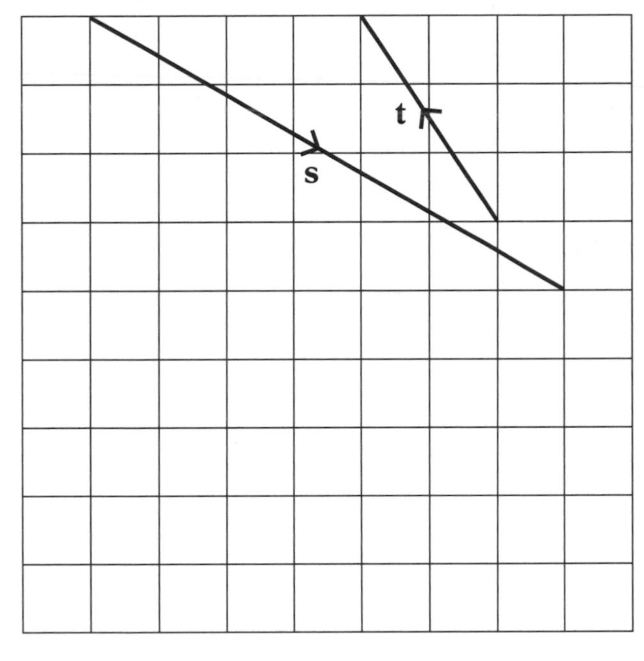

Find the components of **s** + **t**. 2

MARKS | DO NOT WRITE IN THIS MARGIN

3. The diagram below shows the graph of $y = -x^2$.

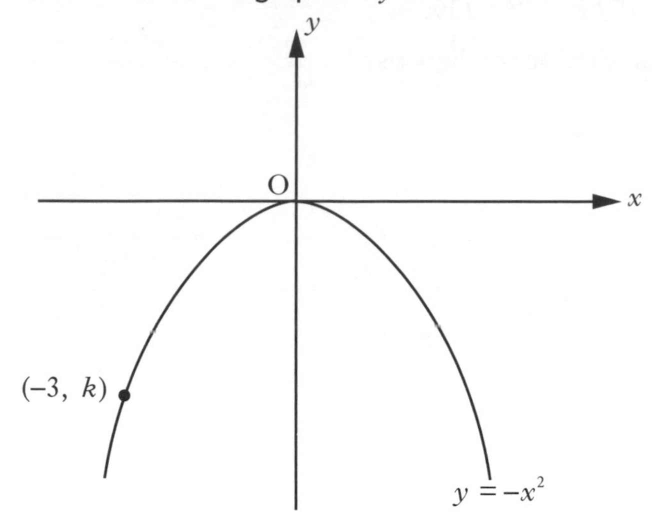

The point $(-3, k)$ lies on the graph.

Find the value of k. **1**

MARKS DO NOT WRITE IN THIS MARGIN

4. A health food shop produces cod liver oil capsules for its customers.

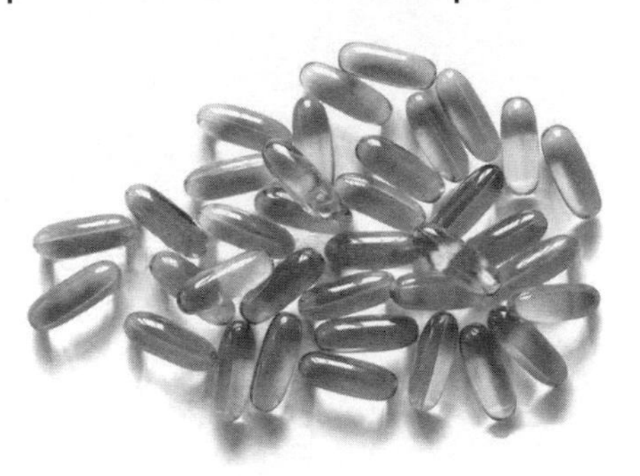

Each capsule is in the shape of a cylinder with hemispherical ends as shown in the diagram below.

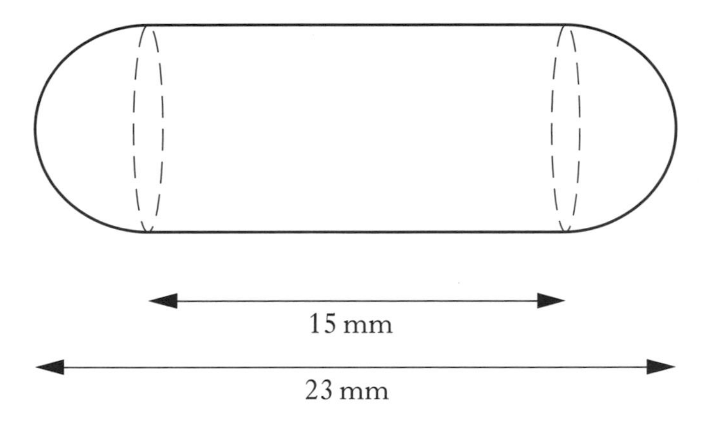

The total length of the capsule is 23 millimetres and the length of the cylinder is 15 millimetres.

Calculate the volume of one cod liver oil capsule. **4**

MARKS | DO NOT WRITE IN THIS MARGIN

5. OABCDEFG is a cube with side 2 units, as shown in the diagram.

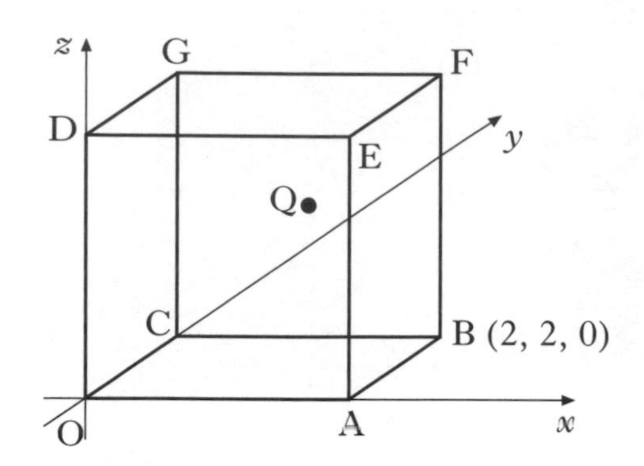

B has coordinates (2, 2, 0).

Q is the midpoint of face CBFG.

Write down the coordinates of G and Q. **2**

6. Express in its simplest form $y^8 \times (y^3)^{-2}$. **2**

7. A straight line is represented by the equation $2y + x = 6$.

(a) Find the gradient of this line. **2**

(b) Write down the coordinates of the point where this line crosses the y-axis. **1**

Total marks 3

MARKS DO NOT WRITE IN THIS MARGIN

8. A pet shop manufactures protective dog collars.

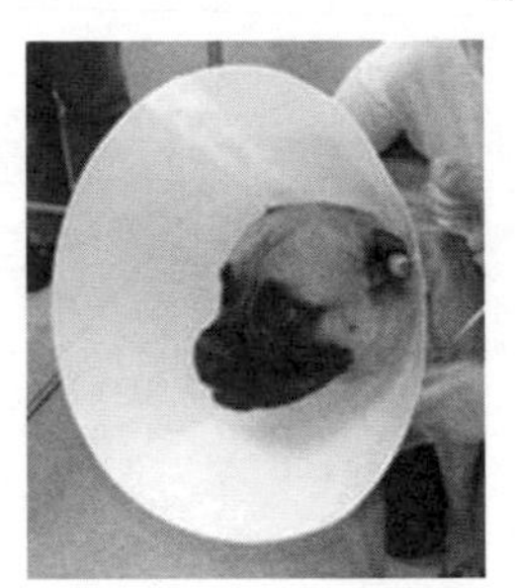

In the diagram below the shaded area represents one of these collars.

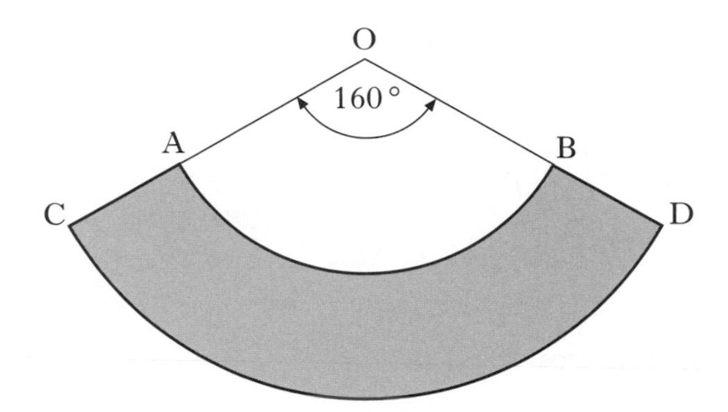

AB and CD are arcs of the circles with centres at O.

The radius, OA, is 10 inches and the radius, OC, is 18 inches.

Angle AOB is 160°.

Calculate the area of the collar. **4**

9. Show that the equation $x(5-2x)=7$ has no real roots. **4**

MARKS DO NOT WRITE IN THIS MARGIN

10. In triangle PQR

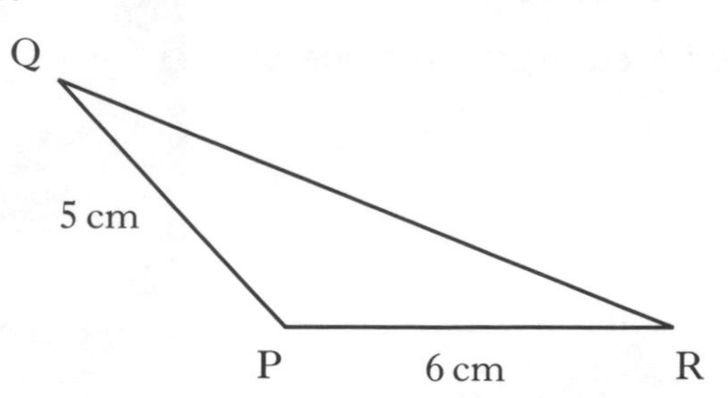

- PQ = 5 centimetres
- PR = 6 centimetres
- Area of triangle PQR = 12 square centimetres
- Angle QPR is **obtuse**.

Calculate the size of angle QPR. **4**

11.

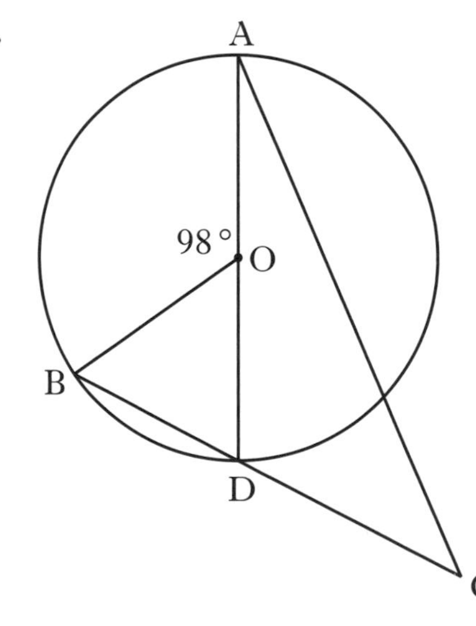

AD is a diameter of a circle, centre O.

B is a point on the circumference of the circle.

The chord BD is extended to a point C, outside the circle.

Angle BOA = 98°.

DC = 9 centimetres.

The radius of the circle is 7 centimetres.

Calculate the length of AC. **5**

MARKS DO NOT WRITE IN THIS MARGIN

12. A right-angled triangle has dimensions, in centimetres, as shown.

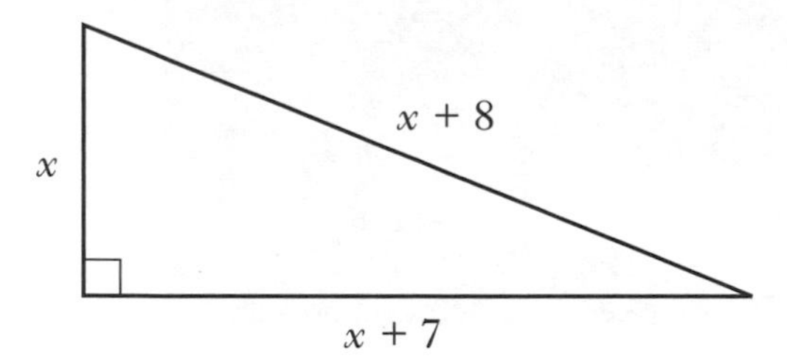

Calculate the value of x. **5**

13. For reasons of safety, a building is supported by two wooden struts, represented by DB and DC in the diagram below.

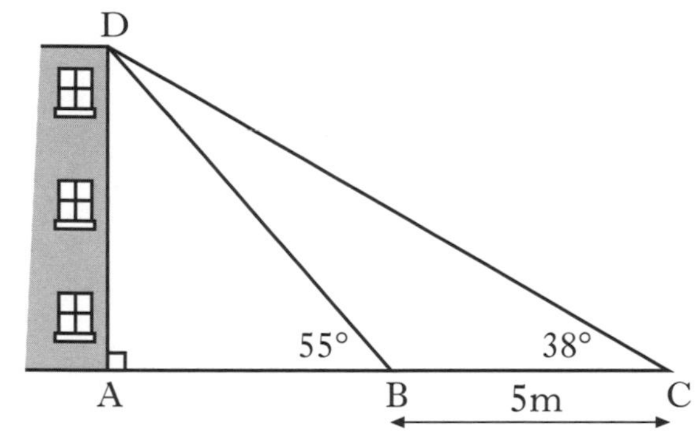

Angle ABD = 55°.

Angle BCD = 38°.

BC is 5 metres.

Calculate the height of the building represented by AD. **5**

MARKS DO NOT WRITE IN THIS MARGIN

14. Due to the threat of global warming, scientists recommended in 2010 that the emissions of greenhouse gases should be reduced by 50% by the year 2050.

The government decided to reduce the emissions of greenhouse gases by 15% **every ten years**, starting in the year 2010.

Will the scientists' recommendations have been achieved by 2050?

You must give a reason for your answer. 4

15. The depth of water, D metres, in a harbour is given by the formula

$$D = 3 + 1{\cdot}75\sin 30h°$$

where h is the number of hours after midnight.

(a) Calculate the depth of the water at 5 am. 2

(b) Calculate the maximum difference in depth of the water in the harbour.

Do not use a trial and improvement method. 2

Total marks 4

[END OF MODEL PRACTICE PAPER]

ADDITIONAL SPACE FOR ANSWERS

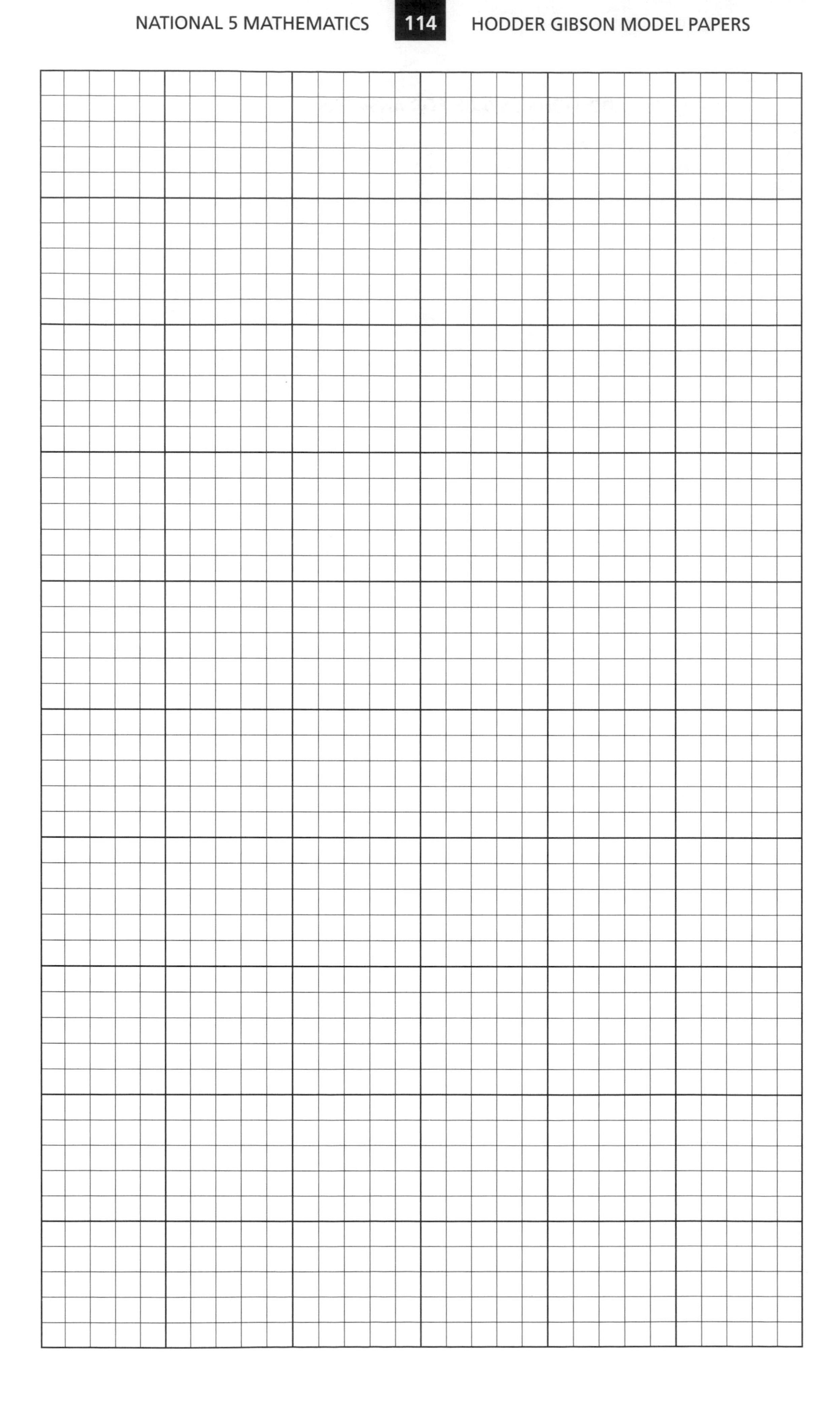

NATIONAL 5

2014

FOR OFFICIAL USE

N5 National Qualifications 2014

Mark

X747/75/01

Mathematics Paper 1 (Non-Calculator)

TUESDAY, 06 MAY

9:00 AM – 10:00 AM

Fill in these boxes and read what is printed below.

Full name of centre

Town

Forename(s)

Surname

Number of seat

Date of birth

Day Month Year

D D M M Y Y

Scottish candidate number

Total marks — 40

Attempt ALL questions.

Write your answers clearly in the spaces provided in this booklet. Additional space for answers is provided at the end of this booklet. If you use this space you must clearly identify the question number you are attempting.

Use **blue** or **black** ink.

You may NOT use a calculator.

Full credit will be given only to solutions which contain appropriate working.

State the units for your answer where appropriate.

Before leaving the examination room you must give this booklet to the Invigilator; if you do not, you may lose all the marks for this paper.

FORMULAE LIST

The roots of $ax^2 + bx + c = 0$ are $x = \frac{-b \pm \sqrt{(b^2 - 4ac)}}{2a}$

Sine rule: $\frac{a}{\sin A} = \frac{b}{\sin B} = \frac{c}{\sin C}$

Cosine rule: $a^2 = b^2 + c^2 - 2bc\cos A$ or $\cos A = \frac{b^2 + c^2 - a^2}{2bc}$

Area of a triangle: $A = \frac{1}{2}ab\sin C$

Volume of a sphere: $V = \frac{4}{3}\pi r^3$

Volume of a cone: $V = \frac{1}{3}\pi r^2 h$

Volume of a pyramid: $V = \frac{1}{3}Ah$

Standard deviation: $s = \sqrt{\frac{\Sigma(x - \overline{x})^2}{n - 1}} = \sqrt{\frac{\Sigma x^2 - (\Sigma x)^2/n}{n - 1}}$, where n is the sample size.

MARKS | DO NOT WRITE IN THIS MARGIN

1. Evaluate $\frac{5}{12} \times 2\frac{2}{9}$.

Give the answer in simplest form. **2**

2. Multiply out the brackets and collect like terms:

$(2x - 5)(3x + 1)$. **2**

[Turn over

MARKS | DO NOT WRITE IN THIS MARGIN

3. Express $x^2 - 14x + 44$ in the form $(x-a)^2 + b$. **2**

4. Find the resultant vector $2\boldsymbol{u} - \boldsymbol{v}$ when $\boldsymbol{u} = \begin{pmatrix} -2 \\ 3 \\ 5 \end{pmatrix}$ and $\boldsymbol{v} = \begin{pmatrix} 0 \\ -4 \\ 7 \end{pmatrix}$.

Express your answer in component form. **2**

MARKS | DO NOT WRITE IN THIS MARGIN

5. In triangle KLM

- KM = 18 centimetres
- sin K = 0·4
- sin L = 0·9

Calculate the length of LM. **3**

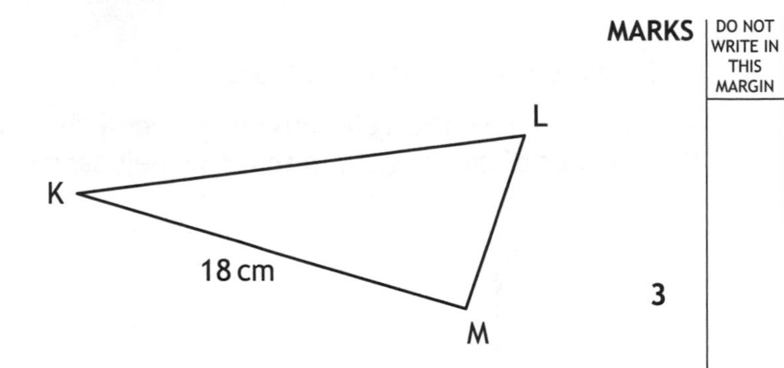

[Turn over

MARKS DO NOT WRITE IN THIS MARGIN

6. McGregor's Burgers sells fast food.

The graph shows the relationship between the amount of fat, F grams, and the number of calories, C, in some of their sandwiches.

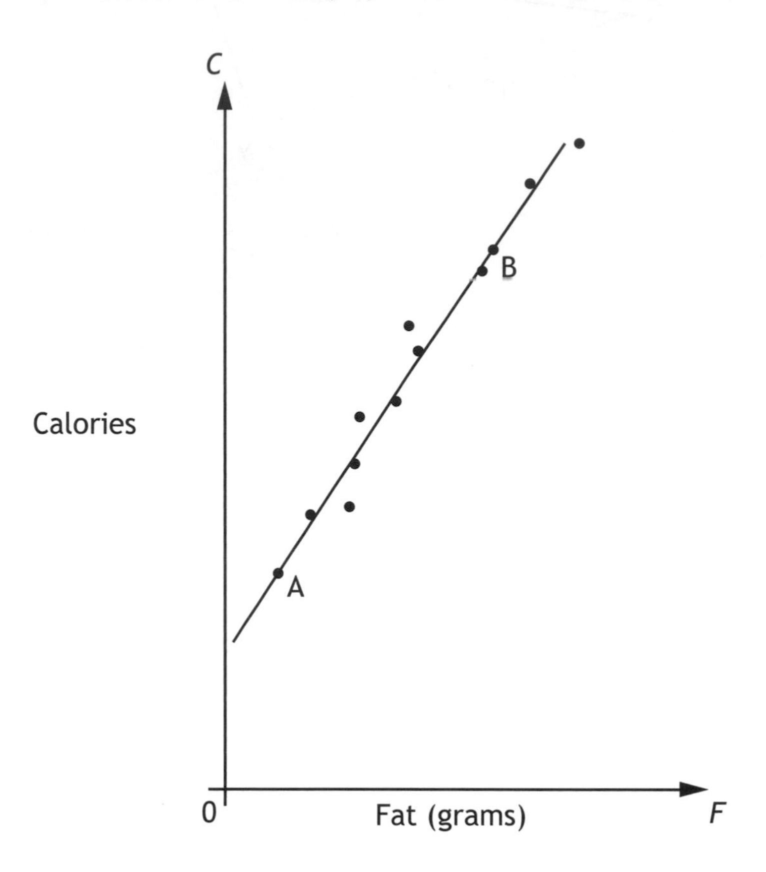

A line of best fit has been drawn.

Point A represents a sandwich which has 5 grams of fat and 200 calories.

Point B represents a sandwich which has 25 grams of fat and 500 calories.

MARKS DO NOT WRITE IN THIS MARGIN

6. (continued)

(a) Find the equation of the line of best fit in terms of F and C. 3

(b) A Super Deluxe sandwich contains 40 grams of fat.

Use your answer to part (a) to estimate the number of calories this sandwich contains.

Show your working. 1

Total marks 4

[Turn over

MARKS DO NOT WRITE IN THIS MARGIN

7. The diagram below shows part of the graph of $y = ax^2$

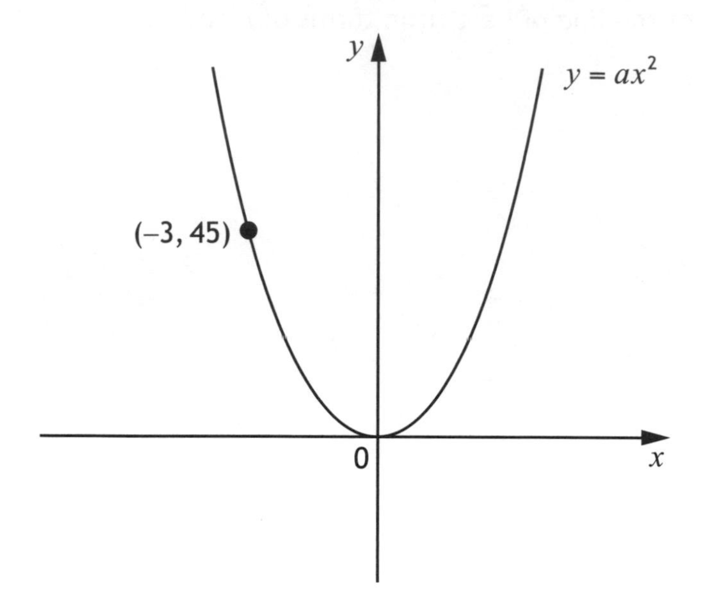

Find the value of a. **2**

MARKS DO NOT WRITE IN THIS MARGIN

8. Express $\sqrt{40} + 4\sqrt{10} + \sqrt{90}$ as a surd in its simplest form. **3**

9. 480 000 tickets were sold for a tennis tournament last year.

This represents 80% of all the available tickets.

Calculate the total number of tickets that were available for this tournament. **3**

[Turn over

MARKS | DO NOT WRITE IN THIS MARGIN

10. The graph of $y = a \sin (x + b)^\circ$, $0 \leq x \leq 360$, is shown below.

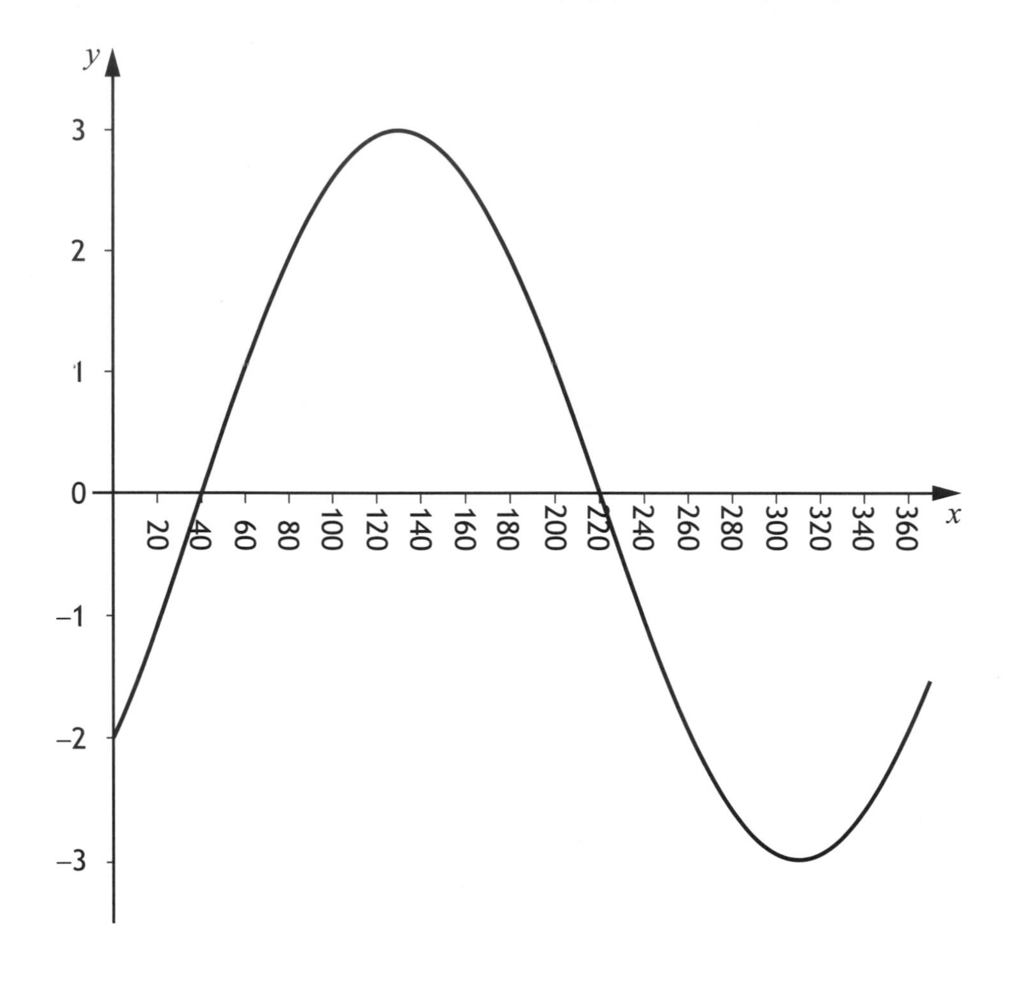

Write down the values of *a* and *b*. **2**

MARKS | DO NOT WRITE IN THIS MARGIN

11. (a) A straight line has equation $4x + 3y = 12$.

Find the gradient of this line. **2**

(b) Find the coordinates of the point where this line crosses the x-axis. **2**

Total marks 4

[Turn over

MARKS | DO NOT WRITE IN THIS MARGIN

12. The diagram below shows a circle, centre C.

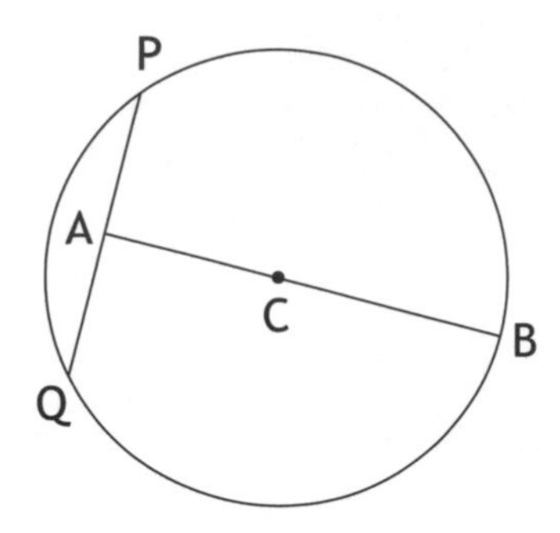

The radius of the circle is 15 centimetres.

A is the mid-point of chord PQ.

The length of AB is 27 centimetres.

Calculate the length of PQ. **4**

MARKS DO NOT WRITE IN THIS MARGIN

13. The diagram below shows the path of a small rocket which is fired into the air. The height, h metres, of the rocket after t seconds is given by

$$h(t) = 16t - t^2$$

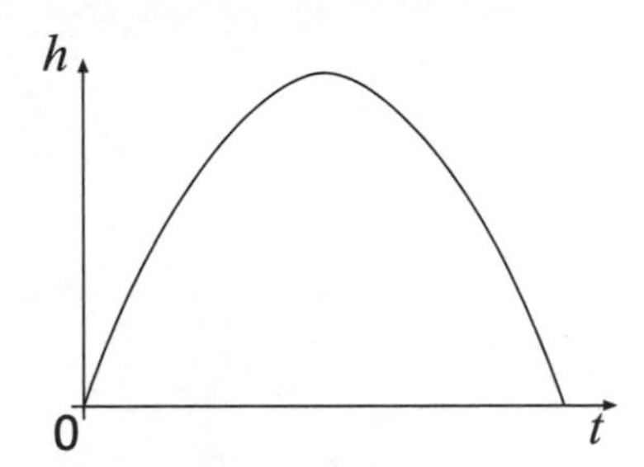

(a) After how many seconds will the rocket first be at a height of 60 metres? **4**

(b) Will the rocket reach a height of 70 metres?
Justify your answer. **3**

Total marks 7

[END OF QUESTION PAPER]

MARKS DO NOT WRITE IN THIS MARGIN

ADDITIONAL SPACE FOR ANSWERS

MARKS | DO NOT WRITE IN THIS MARGIN

ADDITIONAL SPACE FOR ANSWERS

[BLANK PAGE]

DO NOT WRITE ON THIS PAGE

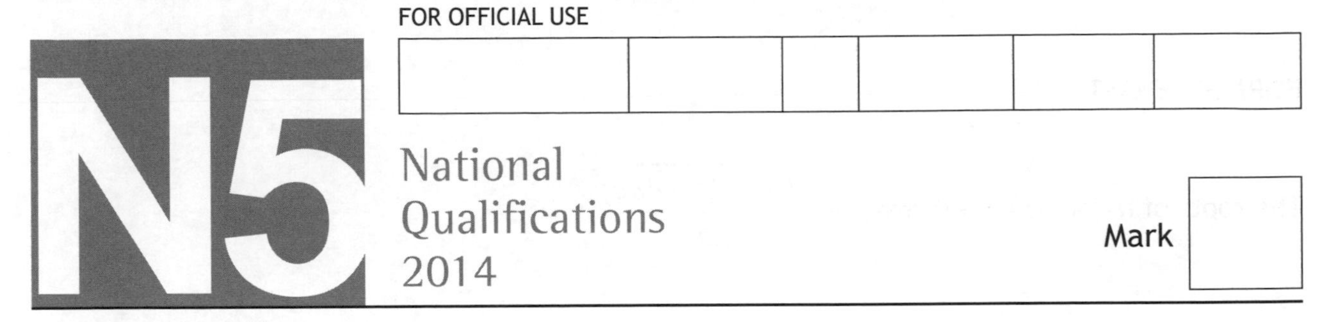

X747/75/02

Mathematics
Paper 2

TUESDAY, 06 MAY
10:20 AM – 11:50 AM

Fill in these boxes and read what is printed below.

Full name of centre

Town

Forename(s)

Surname

Number of seat

Date of birth
Day Month Year
D D M M Y Y

Scottish candidate number

Total marks — 50

Attempt ALL questions.

Write your answers clearly in the spaces provided in this booklet. Additional space for answers is provided at the end of this booklet. If you use this space you must clearly identify the question number you are attempting.

Use **blue** or **black** ink.

You may use a calculator.

Full credit will be given only to solutions which contain appropriate working.

State the units for your answer where appropriate.

Before leaving the examination room you must give this booklet to the Invigilator; if you do not, you may lose all the marks for this paper.

PB

FORMULAE LIST

The roots of $ax^2 + bx + c = 0$ are $x = \dfrac{-b \pm \sqrt{(b^2 - 4ac)}}{2a}$

Sine rule: $\dfrac{a}{\sin A} = \dfrac{b}{\sin B} = \dfrac{c}{\sin C}$

Cosine rule: $a^2 = b^2 + c^2 - 2bc\cos A$ or $\cos A = \dfrac{b^2 + c^2 - a^2}{2bc}$

Area of a triangle: $A = \frac{1}{2}ab\sin C$

Volume of a sphere: $V = \frac{4}{3}\pi r^3$

Volume of a cone: $V = \frac{1}{3}\pi r^2 h$

Volume of a pyramid: $V = \frac{1}{3}Ah$

Standard deviation: $s = \sqrt{\dfrac{\Sigma(x - \bar{x})^2}{n - 1}} = \sqrt{\dfrac{\Sigma x^2 - (\Sigma x)^2/n}{n - 1}}$, where n is the sample size.

MARKS | DO NOT WRITE IN THIS MARGIN

1. There are 964 pupils on the roll of Aberleven High School.

   It is forecast that the roll will decrease by 15% per year.

   What will be the expected roll after 3 years?

   Give your answer to the nearest ten. **3**

[Turn over

MARKS DO NOT WRITE IN THIS MARGIN

2. The diagram shows a cube placed on top of a cuboid, relative to the coordinate axes.

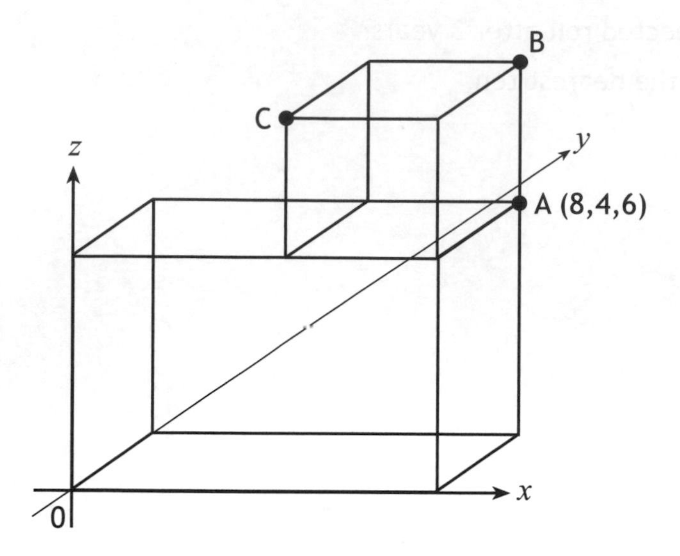

A is the point (8,4,6).

Write down the coordinates of B and C. **2**

MARKS | DO NOT WRITE IN THIS MARGIN

3. Two groups of people go to a theatre.

Bill buys tickets for 5 adults and 3 children.

The total cost of his tickets is £158·25.

(a) Write down an equation to illustrate this information. **1**

(b) Ben buys tickets for 3 adults and 2 children.

The total cost of his tickets is £98.

Write down an equation to illustrate this information. **1**

(c) Calculate the cost of a ticket for an adult and the cost of a ticket for a child. **4**

Total marks 6

[Turn over

MARKS DO NOT WRITE IN THIS MARGIN

4. A runner has recorded her times, in seconds, for six different laps of a running track.

53 57 58 60 55 56

(a) (i) Calculate the mean of these lap times.

Show clearly all your working. 1

(ii) Calculate the standard deviation of these lap times.

Show clearly all your working. 3

MARKS DO NOT WRITE IN THIS MARGIN

4. (continued)

(b) She changes her training routine hoping to improve her consistency.

After this change, she records her times for another six laps.

The mean is 55 seconds and the standard deviation 3·2 seconds.

Has the new training routine improved her consistency?

Give a reason for your answer. 1

Total marks 5

[Turn over

MARKS DO NOT WRITE IN THIS MARGIN

5. A supermarket sells cylindrical cookie jars which are mathematically similar.

The smaller jar has a height of 15 centimetres and a volume of 750 cubic centimetres.

The larger jar has a height of 24 centimetres.

Calculate the volume of the larger jar. **3**

MARKS | DO NOT WRITE IN THIS MARGIN

6. The diagram below shows the position of three towns.

Lowtown is due west of Midtown.

The distance from

- Lowtown to Midtown is 75 kilometres.
- Midtown to Hightown is 110 kilometres.
- Hightown to Lowtown is 85 kilometres.

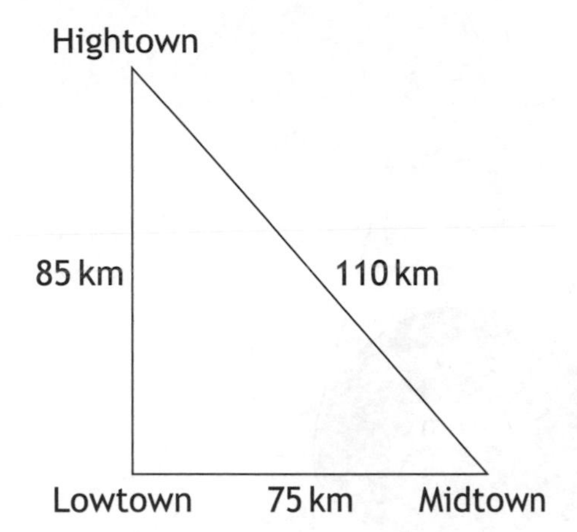

Is Hightown directly north of Lowtown?

Justify your answer. 4

[Turn over

MARKS DO NOT WRITE IN THIS MARGIN

7. An ornament is in the shape of a cone with diameter 8 centimetres and height 15 centimetres.

The bottom contains a hemisphere made of copper with diameter 7·4 centimetres. The rest is made of glass, as shown in the diagram below.

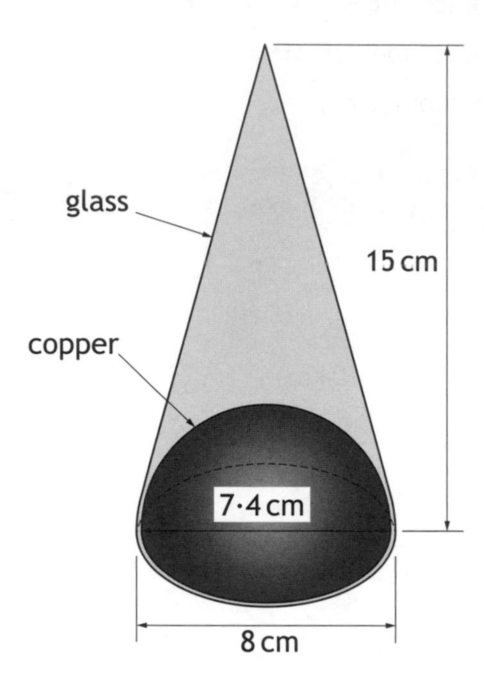

Calculate the volume of the glass part of the ornament.

Give your answer correct to 2 significant figures. **5**

MARKS DO NOT WRITE IN THIS MARGIN

8. Simplify $\frac{n^5 \times 10n}{2n^2}$. **3**

9. Express $\frac{7}{x+5} - \frac{3}{x}$ $x \neq -5,\ x \neq 0$ as a single fraction in its simplest form. **3**

[Turn over

MARKS DO NOT WRITE IN THIS MARGIN

10. In a race, boats sail round three buoys represented by A, B, and C in the diagram below.

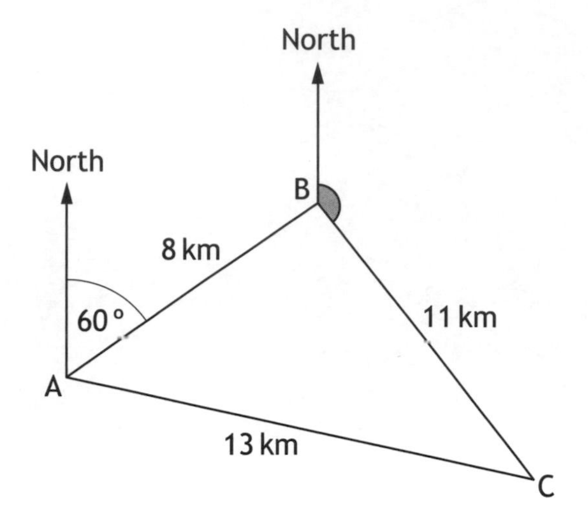

B is 8 kilometres from A on a bearing of 060°.

C is 11 kilometres from B.

A is 13 kilometres from C.

(a) Calculate the size of angle ABC. **3**

(b) Hence find the size of the shaded angle. **2**

Total marks 5

MARKS | DO NOT WRITE IN THIS MARGIN

11. Change the subject of the formula $s = ut + \frac{1}{2}at^2$ to a. **3**

12. Solve the equation $11\cos x° - 2 = 3$, for $0 \leq x \leq 360$. **3**

[Turn over

MARKS | DO NOT WRITE IN THIS MARGIN

13. The picture shows the entrance to a tunnel which is in the shape of part of a circle.

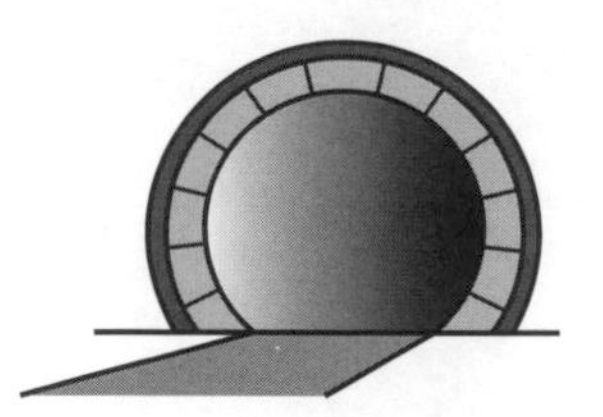

The diagram below represents the cross-section of the tunnel.

- The centre of the circle is O.
- MN is a chord of the circle.
- Angle MON is 50°.
- The radius of the circle is 7 metres.

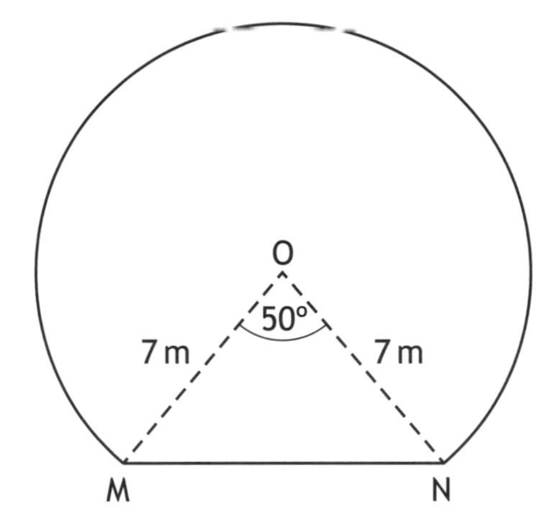

Calculate the area of the cross-section of the tunnel. **5**

[END OF QUESTION PAPER]

MARKS DO NOT WRITE IN THIS MARGIN

ADDITIONAL SPACE FOR ANSWERS

MARKS | DO NOT WRITE IN THIS MARGIN

ADDITIONAL SPACE FOR ANSWERS

NATIONAL 5 | ANSWER SECTION

ANSWER SECTION FOR

SQA AND HODDER GIBSON NATIONAL 5 MATHEMATICS 2014

NATIONAL 5 MATHEMATICS SPECIMEN QUESTION PAPER

Paper 1

1. $7\frac{3}{5}$
2. $2x^3 - 5x^2 - 10x + 3$
3. $7\sqrt{2}$
4. $x = -5$, $x = 1{\cdot}5$
5. $\frac{2\sqrt{6}}{3}$
6. (a) $y = 2x + 1$

   (b) $2 \times 8 + 1 = 17$
7. (a) $x^{-1} + x^0$ or equivalent

   (b) $1\frac{1}{6}$
8. $v = \sqrt{\frac{2p}{m}}$
9. (a) $y = (x - 4)^2 + 3$

   (b)

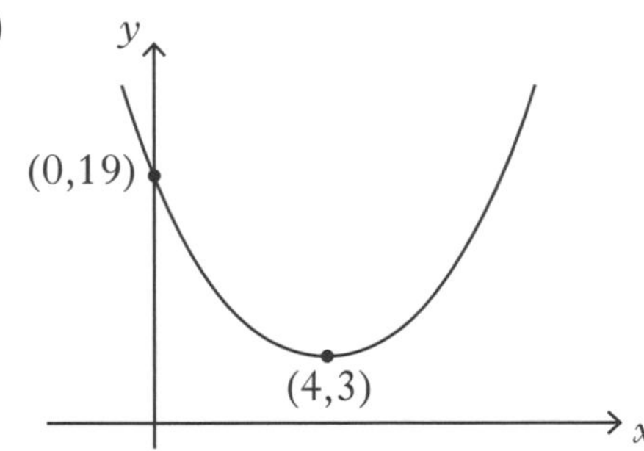

10. (a) $3f + 4r = 185$

    (b) $2f + 3r = 130$

    (c) restricted pass costs £20
    full pass costs £35
11. $\frac{x - 22}{(x + 2)(x - 4)}$
12. (a) $r - 5$

    (b) 10·6 cm

NATIONAL 5 MATHEMATICS SPECIMEN QUESTION PAPER

Paper 2

1. 85·169 miles
2. $1{\cdot}65 \times 10^9$
3. (a) **b** – **a**

   (b) 2(**b** – **a**)
4. –4
5. 9·8 cm
6. 870 cm^3
7. £387·50
8. (a) mean = 21
   standard deviation = 2.1

   (b) On average Machine A packs more sprouts in each bag. The number of sprouts packed in each bag by Machine A is more consistent.
9. 4·1472 litres
10. (a) half of [2 – (–4)]; moved down 1

    (b) 70·5°, 289·5°
11. (a) 1536 cm^2

    (b) 175 cm
12. $p > \frac{1}{3}$
13. (a) 29°

    (b) 249°

NATIONAL 5 MATHEMATICS MODEL PAPER 1

Paper 1

1. $2\frac{5}{6}$
2. $6x^3 - x^2 + 13x - 10$
3. $m = (kL)^2$ **or** $m = k^2L^2$
4. (a) $2\mathbf{u} + \mathbf{v}$

   (b) $\mathbf{v} - \mathbf{u}$
5. 34°
6. $\dfrac{3y}{y + 3}$
7. 27
8. (a) $x = 2$

   (b) 9
9. (4, 5)
10. $a = 5$, $b = 4$
11. $\cos B = \frac{3^2 + 6^2 - 5^2}{2 \times 3 \times 6} = \frac{20}{36} = \frac{5}{9}$
12. $6\sqrt{2}$
13. $\dfrac{7m + 3}{m(m + 1)}$
14. discriminant = 24; roots are real since discriminant > 0 and irrational since discriminant is not a perfect squrare.

NATIONAL 5 MATHEMATICS MODEL PAPER 1

Paper 2

1. £684·70
2. (a) mean = 7, standard deviation = 3·96

   (b) Under the new coach, the team scores more points and is more consistent.
3. M(0, 1, 0), N(4, 2, 2)
4. $y = 3x - 1$
5. arc AB = 0·88 metres; this is less than 0·9 metres, so the staircase will not pass the safety regulations.
6. 5400 cm^3
7. £860
8. 18·3 metres
9. 15 cm^2
10. 1503·5 cm^2
11. (a) −3

    (b) 11·5°, 168·5°
12. 0·35 metres
13. (a) $x^2 + 5 = 3(2x)$

    $x^2 + 5 = 6x$

    $x^2 - 6x + 5 = 0$

    (b) 5

NATIONAL 5 MATHEMATICS MODEL PAPER 2

Paper 1

1. $\frac{4}{11}$
2. $2(m + 3)(m - 3)$
3. -4
4. $\frac{-7}{5}$
5. $4\sqrt{7}$
6. $y = (x + 5)^2 - 8$
7. (a) $24x + 6y = 60$

   (b) $20x + 10y = 40$

   (c) 25
8. $138°$
9. $g = \frac{3}{4}h + 11$
10. $\frac{2}{3}\mathbf{a} + \frac{1}{2}\mathbf{b}$
11. $\frac{\sin B}{4} = \frac{\sin 150°}{10}$

    $\sin B = \frac{4\sin 150°}{10} = \frac{4\sin 30°}{10} = \frac{4 \times \frac{1}{2}}{10} = \frac{2}{10} = \frac{1}{5}$
12. b
13. $\frac{5p}{4}$
14. $\frac{\sin^2 A}{1-\sin^2 A} = \frac{\sin^2 A}{\cos^2 A} = \tan^2 A$

NATIONAL 5 MATHEMATICS MODEL PAPER 2

Paper 2

1. $3{\cdot}24 \times 10^{15}$
2. $x^3 - 2x^2 + x$
3. 2·6 metres
4. $r = \sqrt{\frac{p-q}{2}}$
5. −2·8, 1·3
6. (a) median = 58·5, interquartile range = 22

   (b) In December the marks (on average) are better and less spread out
7. 47·7 kilometres
8. No, as 7754·6 cm^2 $\neq$ 8040 cm^2
9. 7
10. No, as 11700 ($90^2 + 60^2$) $\neq$ 12100 (110^2)
11. 36·4 cm^3
12. (a) $a = -5$, $b = 1$

    (b) P(0, 26), Q(10, 26)
13. (a) (90, 1)

    (b) 48·6°, 131·4°
14. 12 seconds

NATIONAL 5 MATHEMATICS MODEL PAPER 3

Paper 1

1. $3\frac{1}{2}$
2. $x < 1$
3. $(2p + 3)(p - 4)$
4. (a) $S = -4T + 130$

   (b) £10
5. 113·04 cm^3
6. $x = 7$, $y = -2$
7. 540 millilitres
8. (a) $\sqrt{\frac{(1-2)^2 + (1-2)^2 + (1-2)^2 + (2-2)^2 + (5-2)^2}{5-1}} = \sqrt{\frac{12}{4}} = \sqrt{3}$

   or $\sqrt{\frac{(1^2 + 1^2 + 1^2 + 2^2 + 5^2) - \frac{(1+1+1+2+5)^2}{5}}{4}} = \sqrt{\frac{12}{4}} = \sqrt{3}$

   (b) $\sqrt{3}$
9. 750 grams
10. 45°
11. $6\sqrt{2}$
12. (a) $\frac{40}{x}$

    (b) $\frac{40}{x+5}$

    (c) $\frac{200}{x(x+5)}$
13. (a) $(-2, -16)$

    (b) $(6, -16)$

    (c) $y = (x - 14)^2 - 16$

NATIONAL 5 MATHEMATICS MODEL PAPER 3

Paper 2

1. £1 158 000 000 000
2. $\begin{pmatrix} 5 \\ -1 \end{pmatrix}$
3. -9
4. 1022 mm^3
5. G(0, 2, 2), Q(1, 2, 1)
6. y^2
7. (a) $-\frac{1}{2}$

   (b) (0, 3)
8. 313 in^2
9. discriminant = -31; no real roots since discriminant < 0.
10. 126·9°
11. 21 centimetres
12. 5
13. 8·6 metres
14. $0{\cdot}85^4 = 0{\cdot}522$; no, since $0{\cdot}522 > 0{\cdot}5$
15. (a) 3·875 metres

    (b) 3·5 metres

NATIONAL 5 MATHEMATICS 2014

Paper 1

1. $\frac{25}{27}$
2. $6x^2 - 13x - 5$
3. $(x - 7)^2 - 5$
4. $\begin{pmatrix} -4 \\ 10 \\ 3 \end{pmatrix}$
5. 8 cm
6. (a) $C = 15F + 125$

   (b) 725 calories
7. $a = 5$
8. $9\sqrt{10}$
9. 600 000
10. $\boldsymbol{a} = \mathbf{3}$, $\boldsymbol{b} = \mathbf{-40}$
11. (a) gradient = $-\frac{4}{3}$

    (b) (3,0)
12. 18 centimetres
13. (a) 6 seconds

    (b) No, because its maximum height is 64 metres

NATIONAL 5 MATHEMATICS 2014

Paper 2

1. 590
2. B (8, 4, 10), C (4, 0, 10)
3. (a) $5a + 3c = 158{\cdot}25$

   (b) $3a + 2c = 98$

   (c) Adult ticket costs £22·50
   Child tickets costs £15·25
4. (a) (i) $\bar{x} = 56{\cdot}5$
   (ii) $s = 2{\cdot}4$

   (b) No, standard deviation is greater
   or
   No, times are more spread out
5. 3072 cm^3
6. No, with valid reason, eg as the triangle is not right angled since $110^2 \neq 75^2 + 85^2$
7. 150 cm^3
8. $5n^4$
9. $\frac{4x - 15}{x(x + 5)}$
10. (a) 84·8°

    (b) 155·2°
11. $a = \frac{2(s - ut)}{t^2}$
12. $x = 63°, 297°$
13. 151·3 m^2

Acknowledgements

Permission has been sought from all relevant copyright holders and Hodder Gibson is grateful for the use of the following:
Image © JPL Designs/Shutterstock.com (SQP Paper 2 page 12);
Image © Shamleen/Shutterstock.com (SQP Paper 2 page 12).

Hodder Gibson would like to thank the SQA for use of any past exam questions that may have been used in the model papers, whether amended or in original form.